ISLAND TREASURE

The Making and Remaking
Of Petit St. Vincent

LANA SHORE

Copyright © 2015 Lana Shore

Every effort has been made to trace, contact, and cite copyright holders. The author will be pleased to correct any omissions or rectify any mistakes brought to her attention at the earliest opportunity.

All rights reserved. If you would like to use material from the book, other than for review purposes, prior written permission must be obtained by contacting the author at writetolanashore@gmail.com. Thank you for your support of the author's rights.

Library of Congress Control Number: 2015902480

ISBN:10-1505808650
ISBN-13-978-1505808650

With love to my beautiful daughters

∿

Maya, whose love and encouragement is beyond belief;

Rosa, whose devotion and motivation is unsurpassed; and

Morgan, whose patience and virtues are a blessing to us all.

CONTENTS

	Acknowledgments	i
	Foreword	iii
	Introduction	vi
	Jennifer Remembers	xiii
1	The Island of Petit St. Vincent	1
2	Life on Petit St. Vincent: 1700s to 1960s	17
3	How PSV Came to Be	26
4	The Nichols	39
5	Island Construction	62
6	The Backbone of PSV: Haze Richardson and Douglas Terman	73
7	The Men of PSV	87
8	The Women of PSV	103
9	Significant Others	126
10	The Essence of PSV: The Guests	172
11	Island Vegetation and Transportation	187
12	Petit St. Vincent: The Gem of the Caribbean	211
13	The Boats of PSV	233
14	The Remaking of PSV	272
15	Great Losses	287

ACKNOWLEDGMENTS

I would like to offer my heartfelt appreciation to all those who have chosen to contribute to this book and to thank them for their exuberance in relating their stories and memories of the various people and events that made this beautiful island the exquisite place it is. I have met many incredible people throughout this journey and I wish to express my gratitude to each and every one of them for their many contributions which have made this book possible.

I would also like to express my sincere admiration to all the staff members, past and present, who have not been mentioned by name within the pages of this book but who deserve much praise, recognition, and appreciation nonetheless. They were and still are incredibly talented and dedicated and always have exhibited a sense of pride in their work, evident by the exquisiteness which is PSV. I am honored to have met many of these remarkable people.

And last but certainly not least, I am grateful to Phil Stephenson, new owner of this little island in the Caribbean, for giving me the opportunity to write its history which will forever stand as a tribute to all who came before him. Phil has taken on a great task of filling some very important shoes but will create his own incredible history going forward while still upholding the old traditions, and respecting the love and admiration so many already have for PSV. I wish him many blessings for an exciting and remarkable future with the island. I am privileged to have met him and to call him my friend.

FOREWORD

by Sir James Mitchell

After five years in London and traveling over Europe I returned home in January 1967, became Member Of Parliament for the Grenadines and subsequently Minister of Trade, Agriculture, Labour and Tourism.

My responsibility presented the opportunity to grant and administer the tax and duty-free concessions for investment in all the islands Bequia, Mustique, Union Island, Palm (then Prune), and Petit St Vincent, all of which had no jetty, telephones, electricity, airports, hotels or villas.

My own experience developing the family business with Frangipani Hotel along with my exposure in Canada and Europe before politics facilitated an understanding of the investment climate I needed to create and enhance. I was impressed with the early trials to produce fresh water beside the beach in PSV from solar heating and evaporation under plastic sheets.

Our early history of European influence lingered with the boundaries drawn between Grenada and St Vincent separating Petite Martinique and Petit St. Vincent even though so close. I once got a call from Haze one Sunday afternoon to do something about noise pollution blasting away the serenity guests came to PSV to enjoy. I fixed the problem with a call to Grenada's Prime Minister Herbert Blaise in St Georges who was also the MP for the area. PSV is written into Caribbean history with the weekend meeting I hosted there with Prime Ministers John Compton and Eric Gairy of St. Lucia and Grenada when we signed up the first agreement on freedom of movement of citizens between the islands named the Petit St Vincent Initiative. It meant that Petite Martinique workers had unfettered access to the job in PSV ten minutes from home.

I am proud of the quality PSV has created and the continued contribution to life in our islands.

Sir James Mitchell

INTRODUCTION

by Phil Stephenson

Buying an island requires a great deal of vanity; maintaining it afterwards requires just a little insanity.

Friends (and even strangers) often ask why I bought Petit St Vincent in 2010. I jokingly describe it as an impulse purchase made during a period of financial mania, but in truth it was more than that. The combination of the island's location, surroundings, topography, people, and history made it irresistible to me and my two partners Robin Paterson and Colin Hart.

In fact we were buying more than just a piece of property. We did not know it at the time, but we were becoming stewards of the legacy of PSV's previous owner Haze Richardson, guardians of the reputation of its renowned hotel and ensurers of the well-being of PSV's loyal staff.

In November 2010, I sailed with my son Jack from Barbados to visit PSV for the first time after buying the island. On the overnight trip I did not sleep well (and not just because of a two-meter following sea that made my aft cabin into floating version of a child's bouncy castle). Three things concerned me:

1. Would the guests continue to come with the hotel under new management?
2. Would the staff accept me and the changes I would make – or would they strike?
3. Would the planned improvements be able to come in on-time and on-budget?

Ultimately all of these concerned proved unfounded -- except for one. Completing renovation of the hotel on schedule (i.e. by November 2011) required us to "trade money for time" and we spent significantly more than our initial budget. That's what happens when nails are FedExed in order to keep carpenters working.

As for the guests, many were rightfully doubtful that the combination of a Texas-born oil man and a London property developer could bring any improvements to what Haze had done on PSV. After some pretty strong guest feedback, we quickly changed our initial plans to build more hotel rooms (and even private villas) on the island and instead focused on improving the hotel by:

- Renovating all the cottages with custom-made Balinese furniture, new lighting, new bathrooms, air-conditioning, refrigerated minibars, Nespresso coffee makers, Bose iPod stations, and intercoms connected to the main office --- BUT leaving them without TV, phone, or internet service.
- Building a beach bar, boutique, and treetop spa.
- Extending the main dining room to include a lower deck and wine cellar.
- Making an arrangement with Mustique Airways for subsidized private charter service between Barbados and Union that greatly decreased the inconvenience of arriving at PSV for a cost of only US$200 per person more.
- Investing very heavily in infrastructure not seen by guests (and therefore perhaps taken for granted): new generators, water makers, sewage plants, electrical cables, ground lighting, plastic plumbing pipes, and waste disposal facilities.

I'm pleased to say that almost all the repeat guests returned and were unanimous in their surprise and praise for the bulk of what we had done. Some still prefer a few things "the old way," and we try to accommodate these individual desires as PSV has always done.

As for the staff, my concerns about labor unrest also proved unfounded. We undertook to meet with ALL the staff every quarter to explain what we were doing, to ask for their input, and to award cash prizes for outstanding performance. As far as I know, we are completely unique in the Eastern Caribbean in following this practice. As a result of these meetings, we learned a lot and implemented the following improvements for the staff:

- Raised salaries to become the most competitive in the Grenadines.
- Distributed every penny of the 10% service charge paid by guests to all staff.
- Provided hot water in staff accommodations.
- Rebuilt the staff kitchen and canteen and hired better cooks.
- Provided company-paid medical insurance and twice-yearly doctor visits for all.
- Created a facility where employees could borrow no-interest loans to meet medical and housing needs in amounts based on their years of service and pay grade.
- Established a scholarship fund for the children of all employees (which many guests also contribute to; some quite substantially.

One of the greatest compliments I received from staff was being compared to Haze Richardson. Goatie once told me he was sticking around because he saw a little of "the old man" in me. Mattie, who works daily outside the office which was once Haze's and is now mine, remarked a few times on how she'd come in and see me working and think for just a minute it was Haze back again.

That's about the highest praise I could wish for. It's true Haze and I shared some traits. We both were sailors, pilots, navigators, martini-drinkers, and sometimes storytellers. We were both married twice and enjoyed the company of many friends. But these are superficial things.

Haze Richardson was a better man than me in the sense I could never have accomplished what he did. His founding PSV was infinitely harder my renovating it. I merely threw money at challenges whereas Haze met them with physical strength, perseverance, creativity, and courage. I wish I had known him. And, to be honest, I am jealous of him sometimes -- especially when I fail to fix (or even comprehend) the many systems that constantly break on PSV and which Haze understood so completely.

If Haze could ever be jealous of me, it could only be for one thing: that I have a child. My greatest joy is to watch Jack explore PSV, and I hope to someday pass the stewardship of this place on to him, if he agrees. Before then, there are more improvements planned, such as a new scuba diving center being opened by Jean-Michel Cousteau in 2014.

What Haze and I share, in the end, is simply this: a deep and profound love of this island and its people. I commissioned Lana Shore to write a book that could document and reflect that love, and she has done a fine job. Lana lived on the island with her two beautiful daughters while researching this book, staying in the house Haze built for himself. I owe Lana a debt of thanks for her efforts and for putting up with me -- just as she owes debts to those who agreed to speak with her on the record for this book.

I hope you enjoy reading the history of PSV. And know that, whatever else the future may bring, rest assured that we will keep the best of PSV -- its privacy, tranquility, and friendliness -- the way Haze would have wanted it.

Phil Stephenson
January 2015

Phil and Jack Stephenson enjoy boating and a nice cigar

Phil and Jack Stephenson hosting Jean-Michel Cousteau on PSV in January 2014

JENNIFER REMEMBERS

by Jennifer Richardson

There are very few men charming enough to talk a woman into leaving her life and career in the US to move onto a "rock" in the middle of the ocean. Haze Richardson possessed that charm. It wasn't a cultivated charm; it was a natural charm. "Come join me in my adventure," he said. I never hesitated in my decision, I never looked back, . . . **and what an adventure it was!**

Haze Richardson was a very simple man with simple tastes and possessed an absolute rock solid core of decency. From his first involvement with Petit St. Vincent, his whole life was focused on the enhancement and the future of this uninhabited island [if you don't count the original two goats!] He took it into the future without compromising its natural enhancements. His leadership, vision, work ethic, and concern, with not only the project at hand but equally with those he engaged to help him carry out that vision, were major components that helped create what we see today. Under his guidance and steady hand, everyone connected to PSV during his leadership put one hundred percent of their efforts and dedication into making it the best possible place it could be, leaving a wonderful physical plant and legacy for others to pick up and carry forth into the future.

It is with immense pride and great sense of personal satisfaction that I see how PSV is moving forward, achieving new and innovative additions and amenities. I applaud the new owners, their vision, and what they have done to the island to enhance it for all who are lucky enough to discover a sliver of paradise does exist on a one hundred thirteen acre island set off on its own in the Grenadines of the West Indies.

Jennifer Richardson

Jennifer and Haze Richardson

CHAPTER ONE

THE ISLAND OF PETIT ST. VINCENT

It is said that as God pressed his palms down into the warm waters of the Caribbean, the Grenadines were created. Basking along the curved volcanic ridge, reaching from South America to the Virgin Islands lays a rare gift of nature, richly unspoiled and privately owned.[1] This island is known as Petit St. Vincent.

Searching the vast array of documents, periodicals, books and internet articles, very little is mentioned about Petit St Vincent. Now is the time to put this island in the history books. Within these pages we will explore the prehistory of the island, its original inhabitants, and how it became the resort it is today. This book will allow you to become more acquainted with Petit St. Vincent, an island many call home.

Petit St. Vincent is an island in the middle of the Grenadines part of a chain of islands seventy-five miles long which become the southeastern boundary of Atlantic Ocean and the Caribbean Sea. The Grenadines extend from St. Vincent in the north to Grenada in the south. PSV, as it is called, is in the Windward Island group of the southern part of the Lesser Antilles. It is mainly volcanic as it lies on the Caribbean Plate, positioned along its eastern most edge. PSV is located at latitude 12°32'N and longitude 61°23'W. It is made up of approximately one hundred thirteen acres and has two hills, one called Marni Hill which is two hundred seventy five feet above sea level and the other called Telescope Hill which is one hundred twenty six feet above sea level. There are almost two kilometers of white sand beaches which surround the island. Because of its geographic placement, the island is blessed with rich volcanic soil and a warm climate.

According to cartographer L.S. de la Rochette, the Windward Islands were named by the Spaniards, French, Dutch, and the Danes because of their more windward location in the Caribbean Sea and the Leeward Islands were named by the British, sometimes being called the Leeward Charibbee Islands or Forward Islands. In his map dated 1784, the Caribbean Sea is referred to as 'The Charibbean Sea.' Stanislas de la Rochette, who drew this detailed chart of the West Indies, was employed by William Faden who was "the most active and important British map publisher of the period between 1773 and 1823, honoured by being appointed Geographer to King George III."[2] This map was very popular when published in Faden's General Atlas at the end of the 1700s.

THE CIBONEY TRIBES

Its history among the islands of St. Vincent and the Grenadines can be traced back to the original inhabitants who appeared around the year 5000BCE. The Ciboney were hunter-gathers from South America who made their way north through these islands. Archaeologists have found chronological evidence of this cultural migration of the Ciboney tribe buried beneath the land in the form of shell and stone-made tools and roughly shaped axes. These stone-age men used these implements over seven thousand years ago to help gather staples in their diet such as berries, fruits, and mollusks.[3]

The Ciboney were driven farther north when the Arawaks moved in around the third century CE,[4] usually in dug-out canoes which were about fifty feet in length. They originated in the Amazon Basin and settled among the various Caribbean islands, pushing the Ciboneys northward. As the Arawaks expanded among the islands, different cultures developed, each being given a distinctive name. The group settling in the Lesser Antilles was given the name Igneri.

The Arawaks were farmers, differing from the hunter-gatherer tribes of the Ciboney. They established a more hierarchical society. They were considered an advanced civilization consisting of matrilineal leaders,[5] class divisions, and villages of individual households. The class structure, decided by birth, consisted of four basic units: the upper class was called the cacique, the middle class referred to the nitaynos or nobles, the commoners were considered the lower class, and the bottom class was named the naborias, which included the unfree or slaves. The individual households could number between one hundred and two hundred in each village. Each of these villages or groups would have their leader appointed, which usually was a male, but could be female, from the cacique. Respectfully treated by the other tribe members, he would take as many as twenty wives. For the most part, however, the rest of the tribesmen practiced monogamy. Their culture dictated that everything as a whole was shared, including food, land, and belongings.[6]

Existing peacefully, the Arawaks grew cassava and maize and also lived on sweet potatoes and other foods such as pineapples and ground nuts. Whichever island they decided to live on, archeologists believe they had an abundance of bone-forming minerals available to them such as fish, seeds, nuts, and beans, as no presence of tooth decay has been found.[7] Because no large game was usually present on the islands, they consumed fish as the basis for protein in their diet. Their weapons were made of wood in the form of spears, bows, and arrows. The men were responsible for acquiring the meat or fish and clearing the land for the construction of houses, which were built mostly of wood. The women however contributed to a major part of each day's work by being responsible for many chores. They not only planted crops but harvested and cooked the meals for the family. The women always allowed the men to eat first, followed by them, and then the children. They also were weavers and made hammocks out of cotton that grew on the islands.

The Arawaks emphasized their religion, as well as divine ancestry, and believed in life after death. The women of the villages were very excited if they were chosen to be one of the cacique's wives because with this came respect and a higher standard of living. However, there was a distinct disadvantage to this marital arrangement. The cacique, before his ultimate demise, chose three of his most favorite wives and they were to accompany him to his final resting place. He and his chosen wives were buried, burned, or put into a cave. The women were given water and some cassava to take with them on their journey to the Coyaba, or heavenly next world, where it is believed they would continue to live together as husband and wives.[8] Another one of their ritualistic beliefs was they trusted in two Gods: one was a God of the Sky and the other was a Goddess of the Earth. But they also found they wanted daily contact with a spirit which was closer to them. They satisfied their need by naming this guiding spirit, Zemi. It was believed that Zemi could exist as either a human or an animal. This spirit offered protection from severe elements of nature such as fire, brutal winds, destructive hurricanes, and sickness.[9]

They contributed differently to the region from the Ciboney tribe in two other ways. They brought with them horticulture and pottery skills from South America. Their pottery talents are considered by archeologists to be extremely advanced.[10] Evidence of this stoneware survives today on the neighboring islands of Carriacou and Bequia. Sir James Mitchell, former Prime Minister of St. Vincent and the Grenadines, has a beautiful collection of ancient pottery at his famous Frangipani Hotel in Bequia. Besides their stoneware-making skills, they were considered excellent craftsmen in hand-building canoes. They had an abundant supply of whatever they needed to exist and were believed to have been content living a peaceful existence until about the year 1000CE.

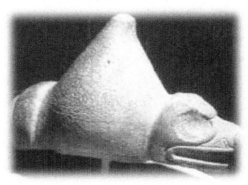

Artifact Example of Zemi

THE CARIB PEOPLE

Following the Arawaks was a tribe descended from another South American group. They were a wandering culture that invaded the Arawak inhabited islands around the year 1000CE. They were known as the Kalina or more simply referred to as Caribs, named by Europeans later.[11] They arrived from the Guianas and Central Brazil in hand-built canoes made out of silk cotton trees. Some canoes measured as long as fifty feet.

The Caribs were extremely different from the Arawaks. They were a mobile, warlike people who lived in a male-dominated polygamous civilization.[12] They became one of the most controlling groups within the islands. In fact, the word 'Caribbean' was named after the Carib people by the Spanish explorers.[13]

The small Carib social system consisted of a war chief [called the Ubutu], priests, elders, warriors, and hunters. All decisions within the village were to be made only by men. The Ubutu was the chief who had absolute power over everyone in his tribe. He was not born into his role as chief through inheritance but rather was elected by the elders of the village. He was chosen based on his ability to lead a successful raid and had to possess the attributes of being physically strong, extremely brave, and proficient in battle. When the Ubutu was selected, his responsibilities included deciding which enemy to attack, planning and implementing a raid, and awarding the spoils of the raid to the tribe members he decided were most brave. He also chose who was going to skipper the piragua, or canoe, that was built for the raid.[14]

There was another ruling person who led in times of peace. He was called the tiubutuli hauthe. This headman's responsibilities included making laws and supervising the daily farming and fishing chores. Outside of these particular duties, this headman had almost no other authority.

The elders and priests of each village were highly admired and well respected. These elders had each been a past warrior and therefore were able to know what qualities were essential for the next Ubutu. These elders did not need to do much on a daily basis as they held a place of reverence among their families and were therefore well taken care of. A few select tribesmen were chosen to be priests called Boyez. The Boyez were believed to have special powers that could ward off the evil spirits they called Maboya. The Boyez also lead any communal ceremonies and sacrifices which took place.

Each male had many wives and in addition, the Ubutu himself kept a large amount of slave women to be given to prized warriors as gifts for their accomplishments and skill in battle. Although most of the marriages were prearranged, the men were expected to provide a hut and any furnishings for each woman they married. A garter on the leg of a young woman indicated she was available and was worn until she married, which was usually by the age of eighteen. All Carib women however continually wore bands around their ankles and knees so their calves became enlarged which apparently would attract the men.

The women were expected to do a very large share of the work each day. Their duties included taking care of their living quarters, raising and providing care for the children, and cutting and gathering wood for the fire. The women not only planted, cultivated, and harvested the crops but were also responsible for trading their harvest among their village. They grew staples such as corn, sweet potatoes, tomatoes, maize, manioc, and cassava. They cooked the fish and meat the men brought in which consisted of jack fish, conches, mollusks, and cockles caught from the sea as well as tatou [armadillo] and agouti [guinea pig-like rodent] found on the island.

Another very important chore was to weave and hammocks were one of the essential items they made.[15] Archaeologists have discovered evidence of these, some with elaborate designs woven into them, as well as the spinning tools that had been used to construct them. The women also used this skill to weave nets used for fishing and to make traps to catch birds.[16]

Natural resources of the island were vital to the Caribs and were used in many ways. Rugs were made from the local cotton plants and then colored by using natural materials such as tree bark, leaves, or plants boiled to produce a dye. They used bamboo, straw, and bark to weave baskets and mats to sit on, a tradition that is still continued today in the Caribbean. They would also gather together tref leaves to use in bathing to ward off bad luck or spells. They chose to live by the sea since it provided their main source of protein. The shells from their conch fishing would be lined up in very large piles, not unlike the mounds of conch shells we see today in the area of Petit St. Vincent.

When a woman had a male child, he would reside with the mother until the age of four. After that, male children were moved to separate houses where all men of the tribe lived. These separate living quarters were called carbets, different from the huts the women resided in. It was believed that allowing male children to live with their mother past the age of four would weaken their abilities to become warriors. As they grew older, most of the boys were trained to fight and to hunt. Their daily duties included fishing and building the carbets, huts, and canoes.

Death for a tribesman meant an extraordinary voyage. They believed they would either end up in 'heaven' or in their idea of 'hell.' However, if they had been considered one of the bravest of warriors, they would be transported to a 'fortunate island' where Arawak slaves would be to wait on their every need. If, on the other hand, they had been considered a coward within their lifetime as a warrior, they were transported to an island where they themselves would be an Arawak slave.

In either case, whenever a Carib Indian died, there were certain steps that had to be followed. First, they were carefully examined to see if sorcery had played a role in their downfall. Second, their body was washed and painted with a red dye and their hair combed with oil. The body was then put in the ground not lying down but rather sitting up on a stool. In order to keep the body warm, a fire was lit close to the opening, and food and water was offered by the relatives for a period lasting ten days. At the end of this time, the body was covered with dirt and the deceased's possessions and home were burned to the ground.

The Caribs were considered a fierce tribe and over the years pushed the Arawaks northward or killed them, except for the Arawak women they kept as slaves. They were very proud of the fact they could withstand enemies, as well as swim and fight extremely well. When the Caribs needed more women and food, they would travel to the islands farther north and raid the communities, killing Arawak men as they went. They used wood, bone, and stone to fashion into weapons such as bows, arrows, and pointed spears. They would harden these by heating them in a fire and then poison the point of the arrows and spears with sap from the local manchineel trees. The Caribs would also tie an enemy to the trunk of this tree and the sap would drip on the body, making for a very slow, painful death. This same tree's leaves were sometimes used to contaminate the enemy's water supply.[17] Supposedly, arrowroot was made into a poultice which the Arawak's could use as a possible remedy.[18]

Such was the way of life for the Caribs in the Lesser Antilles. The Arawaks continued to live peacefully, for the most part, in the far northern islands of the Greater Antilles, where they had been driven. Their lives continued this way until the next foreigners arrived, not coming from South America, but rather from Spain. It was Christopher Columbus, who arrived in 1492. As most of us realize, Columbus believed he had sailed to the East Indies. Thus, the original inhabitants are misnamed Indians and later, Europe regards the entire region as the West Indies.[19] Also, because of Columbus' mistake, he called what is now the Caribbean Sea, the Sea of the Antilles or Antilia, an altered form of Atlantis, the mythical lost continent.

Columbus' first encounter was with the Arawaks who were living in the area. He found them to be very obliging and helpful. They offered him most anything he and his men wanted as was true to their nature. In his writings, Columbus described them as being short with a medium build, weaker than Europeans or Africans, having brown skin and long, coarse, straight hair. They did not wear much clothing and only small gold ornaments adorning their nostrils. This small decoration would later be the downfall of their civilization. Little did the Arawaks know that Columbus would not be friendly and he and his men were after gold which was part of their body's embellishments.

Columbus wrote of the Arawaks:

"They . . . brought us parrots and balls of cotton and spears and many other things, which they exchanged for the glass beads and hawks' bells. They willingly traded

everything they owned. They were well-built, with good bodies and handsome features . . . They do not bear arms, and do not know them, for I showed them a sword, they took it by the edge and cut themselves out of ignorance. They have no iron. Their spears are made of cane . . . They would make fine servants . . . With fifty men we could subjugate them all and make them do whatever we want . . . As soon as I arrived in the Indies, on the first Island which I found, I took some of the natives by force in order that they might learn and might give me information of whatever there is in these parts." [20]

Columbus decided to make the Arawaks into slaves and shipped them off to Spain as trade for cattle from King Ferdinand and Queen Isabella. However, he told the King and Queen these men were cannibals, hoping the rulers would take them as slaves. The people he sent to Spain were Arawaks, not the Caribs who were at one time rumored to be cannibals. The Arawak name for the Caribs was Canibas, and because of their fierce warlike ways, they were imagined to be flesh eaters. However, this myth has been dismissed by anthropologists who feel rather they used their enemies in ritualistic ways. Sometimes they would wear the enemy's teeth around their neck as a sign of triumph and only ate them in rare instances when they might have felt they could obtain the bravery of the enemy by doing so. 'Cannibalism' was originated by Columbus to further his agenda of enslavement.[21]

Many of these Arawaks were loaded onto the *Pinta and Niña*, two of Columbus' sailing vessels, and were taken against their will to Spain while others were taken to Hispaniola. Many of the slaves died tragically when the weather turned cold and from lack of nutrition. By the year 1500CE, over forty thousand of the Arawak Indians had been taken against their will and forced into slavery.[22] The Spanish proved too much for these welcoming Arawaks. They had more military, weapons, and soldiers than did the natives as well as guns and dogs. The enslaved were worked to death, died from starvation or diseases the Spanish brought to their lands, or decided that suicide was better than living the life of a tortured slave.[23]

Columbus returned on a number of voyages and in 1498, on his third expedition, he discovered St. Vincent. The Carib inhabitants who occupied the island called it Hairoun, meaning Land of the Blessed. [Interestingly, the country's beer bears the same name.] Because they were a much fiercer tribe than the Arawaks, they were able to stop the Europeans from taking over. At the same time, the Caribs found that some of the island slaves around Barbados were escaping from their masters in canoes. Originally they would make them return but discovered quickly they could be a help rather than a hindrance in their quest against the French and Spanish troops. It has been estimated that by the 1670s, five hundred Barbadian slaves were on the island of St. Vincent[24] and their combined efforts with the Caribs enabled them to hold off Spanish invasion.

Much was written by the Spanish and then subsequently the English and French about the Caribs. They were referred to as *les sauvages*, people who "prefer to die of hunger rather than live as a slave, proud, exceedingly vindictive" and people who should not be offended.[25] Samuel Eliot Morrison, a Harvard historian, wrote about Columbus' enslavement of the Arawaks: "The cruel policy initiated by Columbus and pursued by his successors resulted in complete genocide."[26]

Two royal Spanish proclamations dated 1508 and 1512 allowed for the capture and enslavement of Caribs on various islands including St. Vincent and Grenada. By the end of the 1500s however, the Spanish realized there was no gold to be had in the Lesser Antilles and were confronted by numerous fatalities when fighting the Caribs. The Spanish therefore decided to focus their efforts in the Greater Antilles. The division of areas was referred to as the "poison arrow curtain."[27]

In 1675, a Dutch ship was wrecked between St. Vincent and Bequia and lost many of the people on board, but some of the slaves survived. The Caribs allowed them to settle on St. Vincent, adding to their stronghold. The subsequent mixed race which followed their interbreeding with the Caribs was self-named the Garifuna, or Black Caribs. Garifuna means 'cassava eating people.' The Black Caribs and the original Caribs, to be later called the Red or Yellow Caribs, lived peacefully together. As a whole, they were able to hold off offending attackers until the 1700s. By then, the Garifuna outnumbered the original tribe. Unfortunately, when French troops later took over the island, many of the Garifuna tribesmen were put on a ship to Belize to be used as slaves.

Major changes were about to happen in 1667 in St Vincent. Sir Thomas Modyford, former governor of Jamaica, maintained that the Caribs must be defeated at all costs. He justified this by saying they were resisting Christianity. The English's first effort would be to systematically pursue holdings in the Windward Islands through the Willoughby Initiative of 1667. William Lord Willoughby, who was governor of Barbados at that time, realized there would be great profit in making this area a slave-based sugar plantation.

For about seventy years, until 1700, colonization was being hard-pressed in both the Leeward and Windward Islands but the Caribs presented a strong wall of resistance. By then, England, France, and Spain each believed they had claim to the Caribbean. They all thought great wealth could be acquired, so these powers scrambled for ownership of the area, unfortunately "with disastrous results for the indigenous people."[28]

During these years, the English and French were both setting up plantation settlements and thought they had exclusive rights to the islands. This matter of ownership was finally resolved in the Treaty of Utrecht of 1713, ending the War of Spanish Succession, which gave the English full control. However, the Caribs were holding strong the islands of St. Vincent, Dominica, St. Lucia, Tobago, and Grenada. "While other native Caribbean people suffered large scale slavery at the hands of the Europeans, the Caribs were never found in large numbers working the mines, on latifundia, or on plantations in the Lesser Antilles."[29]

By 1722, French settlers had finally arranged peaceful cohabitation with the Caribs. They were growing crops to sell such as cotton, sugar, indigo, and tobacco. Although Britain tried at that time to secure the island of St. Vincent, the French retained control. For many more years, possession of the islands continued to change. Between 1739 and 1748, wars erupted between the British, Spain, and

France ending with the Treaty of Aix-La-Chapelle being signed in 1748 which pronounced St. Vincent, Dominica, St. Lucia, and Tobago neutral islands.

In 1759, because the British were still contesting French presence on the mainland of America, the Prime Minister decided it would be best to have some bargaining power over them. It was therefore ordered that the French islands of the West Indies be captured. During the ensuing peace agreement of 1761, the British ordered the islands to be traded to the French for the whole of Canada.[30]

In the first Treaty of Paris in 1763, the 9th Article gave Britain the French colonies of St. Vincent and the Grenadines, among others. From 1769-1773, the First Carib War broke out between the Caribs and the British. There was no agreement that could be reached so ownership remained unresolved.

In the following letter, dated 1778, a copy of which was sent to the future owner of PSV [Willis Nichols], a recommendation for Grenadines' fortification from 1778 for the island of Petit St. Vincent is documented:

"In 1778, a British engineer named [Robert] Morse went through the Grenadines with an eye to preparing their defense where such appeared responsible against French and American attackers. Lord Macaringy, Governor of Grenada, skeptically since he thought ships superior to land defenses, forwarded them to Lord George German. The originals are held in P.R.O. in London, reference is co-101-22 pp 70-81."[31]

The letter reads as follows:

Petit St. Vincent is about two leagues [about 7 miles] South East from the Union, It is the property of Mr. Thomas Bennett of Grenada and contains by Computation One hundred and Fifty acres. There are at present only one white Man and Ten Slaves, A number having been withdrawn on account of American Cavaliers – Fifteen thousand weight of cotton has been produced here in a Year, but the Crops of late have been diminished greatly -- The surface of this Island is tolerably good, but the Soil poor. The same mistake has been committed here, as in most of the other Grenadines, That of clearing away too much of the Wood, and not leaving sufficient shelter for the Cotton – There are no Defenses [sic].[32]

During the American War of Independence in 1779, France again regained control of the island of St. Vincent, with the aid of the north-occupying Black Caribs. But again in 1783, the second Treaty of Paris restored control of the island to Britain. The years 1795 through 1797 saw the Second Carib War resulting in the Caribs gaining control of St. Vincent. In 1797, the commander in chief of the British forces in the West Indies, named Sir Ralph Abercromby, fought the conclusive battle against the Caribs which ended their hundreds of years of opposition to the colonization of the Europeans. Within this enormous battle, the chief of the Garifuna, Joseph Chatoyer, was killed and over five thousand Caribs were sent to Roatán, an island off of what is now Honduras. The Caribs were basically eliminated by this 1797 battle.

Unfortunately during this time, slavery continued to run rampant among the islands and many accounts have been recorded of the unspeakable living and working conditions the men had to sustain. Historically women were not discussed as much but their lives were as tragic as the men since they were enslaved as well. In the book *Engendering History*, a woman passed by a slave market in 1770 and said she "was brought to see [slaves] as subhuman brutes outside her frame of reference and therefore unworthy of pity, yet alone empathy."[33] Sadly, this is how many men and women were viewed during these horrific times.

Slavery continued until 1833 in the colonies within the Greater and Lesser Antilles. Up until then, according to the British National Archives, approximately 2.2 million slaves had been taken to the Caribbean from Africa. This does not include the thousands of Arawak and Carib Indians already enslaved. Emancipation Day is considered to be July 31, 1834 within the British West Indies but not until 1838 did everyone truly become free. The law was written that only children under age six were immediately freed but it could take as long as four years for adult slaves to be released because under the Emancipation Act, some slaves would continue to be bound to their masters for a certain period of time.[34]

Once British rule was firmly established, St. Vincent went through periods of colony status. For example, until 1959 it was part of the Windward Islands Federation, and concurrently between 1958 and 1962, it was also considered a part of the West Indies Federation. In 1969, St. Vincent and the Grenadines became a British Associated State, which allowed for a self-governing system. And finally, in 1979 a referendum was passed allowing for its full independence, the last of the Windward Islands to attain sovereignty.[35]

CHAPTER TWO

LIFE ON PETIT ST. VINCENT
1700s TO THE 1960s

In the late 1700s, Thomas Bennett and ten slaves inhabited PSV, as documented in Robert Morse's letter, written in Chapter One. The next recorded inhabitant of the island in the early 1800s is Richard Bethel. His descendants reside today on the neighboring island of Petite Martinique which, although lies in close proximity to PSV, is actually ruled by Grenada and is not under St. Vincent authority as is Petit St. Vincent.

Richard and his wife, Mary Ann Constance St. Hilaire, along with others, lived full time on Petit St. Vincent. It is unknown by the current inhabitants of Petite Martinique exactly which year that would have been. However, Richard and Mary Ann had many children and grandchildren who were born on the island. Many were also buried on the island.[1] They had small homes where they grew vegetables and raised goats. Richard and Mary Ann Bethel were considered at that time the owners of the island.

According to the Bethel's Catholic beliefs, when a person died, it was essential for a priest to oversee the burial so that a direct means of going to heaven would be guaranteed. At that time, there was a Catholic priest named Father Joseph Modeste Aquart, who occasionally provided Sunday mass on Petit St. Vincent. He was based at the St. Patrick Catholic Church on the nearby island of Carriacou and would travel between the islands of Union, Canouan, Mayreau, Petite Martinique, and PSV to provide mass. He served as priest from the latter years of the 1890s until 1924.

Richard and Mary Ann wanted a guarantee they would go to heaven so they decided the best way to assure their accession was to give the Catholic Church half ownership of the island! It was the agreement that no matter where he happened to be or who he was visiting Father Aquart would immediately come to Petit St. Vincent and oversee the burial of Richard or Mary Ann Bethel.

Father Aquart, who was also referred to as Father Acar by the inhabitants of Petite Martinique, accepted this proposal. Being half owner of the island, he decided the first order of business would be to raise animals on the land. It was arranged they would be cared for by the Bethels while he was travelling from island to island on church duties. In their care he left two donkeys, four head of cattle, one pregnant horse, fifty sheep, and an unknown number of goats and chickens. After being gone for awhile, upon his return, he found all the animals had perished. It is believed the animals had no water.

The people on PSV were usually able to get water from a pond that still exists there today even though they had suffered a period of drought around that time. It is thought there was not enough water for both people and animals so the animals perished. Father Acar was extremely distressed especially about losing his horse, which was one of his most prized possessions and upsetting a priest was a very bad omen, the Bethel's believed. Therefore, in an attempt to appease him and avoid a worse fate, Mary Ann decided to give the other half of the island to the church. She bequeathed Petit St. Vincent to Reverend Joseph Modeste Aquart in her will dated May 11, 1876.

Amazingly, after all the effort the Bethels put into the 'guarantee' of heaven, Father Aquart never presided at either of their burials. Mary Ann passed away in March 1884 and it is unknown when Richard passed. Because the Father was not there, what happened to them after death will always spiritually remain a mystery! Both were laid to rest on the island of Petite Martinique.

In a letter dated 1912, residents of Petite Martinique wrote to Archbishop John Pius Dowling in Trinidad. This shows that many other residents met with the same fate as the Bethels because Father Aquart was not always able to be present at burials:

"We Pray that your Grace may give us a residential Parish Priest. We do not mean that Father Aquart neglects us in any manner but having much to contend with his principal

parish he is unable to devote much time to us and as far as our sister islands are concerned [Mayreau and Canouan] he had hardly find time to visit them. Many there are who pass out of their mortal life without seeing the priest . . . Father Aquart erected a presbytery here so that someday a residential priest may be stationed in this island. . .[sic]" [2]

Around the year 1913, the Catholic Church decided to lease the island to Arthur Ollivierre, according to his great granddaughter Sherron Roberts.[3] He owned a boat named *Geraldine Louise*. Under the lease, there were two stipulations: The first was that no valuable trees, such as cedar, could be cut but that the mangrove and manchineel trees should be taken down. The second stipulation was that the church must offer the island to Mr. Ollivierre at the end of the lease if it was decided to be sold.

Arthur Ollivierre was originally from Bequia. His parents came from France and were whalers. He married a lady named Mary Roberts, who was from Petite Martinique and one of eleven children. Arthur and Mary had four children of their own named Adelaine, Lilian [Lily], Eva, and Arthur II [Cyprian]. While he was leasing the island from the church, Arthur used the timber on Petit St. Vincent to earn money by taking cut trees to Barbados. He also raised goats for meat and allowed Petite Martinique residents to pick up firewood. Each day Arthur would go to PSV in his rowboat from Petite Martinique.

In an account by Doradine Ollivierre, who was married to Arthur's grandson, Fred Ollivierre, the people who lived on PSV before it was leased to Arthur had water from the island's pond to drink, had raised goats, and had planted potatoes. She also told the following story about Arthur, who died a terrible death:

There was a building close to where he lived on Petite Martinique where cotton was stored while it was waiting to be sold. It was called The Cotton House. "The building was also used as a hall for dances and parties. One night during one of the parties, a fight arose. Arthur went to try to stop the fight and he fell down on a bottle and was cut. His wound later became infected and he died."[4]

Tragically, Arthur passed away from a tetanus infection at age forty four. [The tetanus vaccine was not widely used until World War II, even though it was first discovered in 1924.[5]]

Following his death, a few events happened within a short time which changed the course of PSV's history: Father Aquart died in 1920 and the lease on the land was going to expire. It was then that the Catholic Church came to Mary advising her of their decision to sell the island. The church asked if she would like to buy it from them, honoring their original agreement with her late husband, Arthur. Mary decided that she would indeed purchase Petit St. Vincent.

Petit St. Vincent was sold in its entirety to Mary Ollivierre and her daughter, Geraldine Bethel [nee Ollivierre, otherwise known by the family as Lillian or Lily] on December 22, 1926. The total cost was £300 with Mary and Lily each paying half. The fact that young Lily was now part owner of the island was not known to all of Lily's siblings until the research on this book was underway and it was discovered through viewing the sale documents. According to Ms. Roberts, "the culture of Petite Martinique on the death of the parents is that the youngest son is the one who inherits."[6] Apparently, Arthur and Mary's only son, Cyprian, had been the one thought to have begun purchasing the island from the church. According to the family, he had the first four receipts made out in his name and he continued to send money to Mary to make payments on the island while away working in British Guyana. Cyprian's family believes his mother Mary had used the funds that he sent her for the purchase of the island but subsequently the island's sale documents were eventually put only into her and Lily's name.[7]

Although Mary and Lily lived on Petite Martinique, they went to Petit St. Vincent often. Lily was mostly in charge of taking care of chores on the island. She had two ponds dug for the collection of water for animals and to furnish water to Petite Martinique residents during severe dry seasons. Lily had a vegetable garden and raised goats. It is believed she did not sell this food but would use it to feed her family. Sometimes the goats were found to be missing and were believed to have been stolen by residents of Union Island.[8] There was a small boat house that was standing on the island where she could stay if she wanted, but Lily usually went back and forth daily from her home in Petite Martinique.

As the years went by, Lily also had the added responsibility of taking care of Mary when she began suffering from arthritis. Lily took care of her until Mary's death in March of 1955. It is said that prior to Mary's passing, they went to St. Vincent and had the entire island's ownership transferred over to Lily's name.

Lily was married to Peterson Bethel. They had four sons named Ellis, Sandy [known as Antony], Sylvester [known as Eugene], and McKennie [known as Gilbert]. Peterson was a mariner by trade. They lived happily with their children in Petite Martinique until two terrible blows hit the Bethels in 1951. Peterson passed away in the early part of the year. Worse yet, three of Lily and Peterson's sons were killed in a shipwreck just five months later. Sandy, Sylvester, and McKennie were on a cargo ship bound to Dominica from Trinidad. The name of the boat was *Lily Rose* and it encountered a terrible storm. The other boat that was travelling with them but was slightly ahead saw the boat sink but was unable to get back to it to offer any help. There is a gravestone placed near the Bethel's home on Petite Martinique in their honor. Two sons of Lily's sister were also onboard and tragically suffered the same fate.

Gravestone of Bethel boys on Petite Martinique

Ellis Bethel was the only son of Lily and Peterson that was not aboard the *Lily Rose* that tragic night. He married Fedelin Bethel in 1952. She has kindly shared many of these accounts. Because of this shipwreck, Lily was left with three of Sandy's children to raise. Ellis returned to live with Fedelin and Lily to take care of all of them after the accident.

Just four short years later, Lily suffered a third tragedy. The island of Petite Martinique was hit by a devastating hurricane in 1955. According to the *Barbados Museum and Historical Society*, Hurricane Janet went on to become a Category 5 hurricane with winds topping 175 mph.[9] Hurricane Janet directly hit the islands of Carriacou, Grenada, and Petite Martinique. There was a warning broadcast by the Windward Islands Broadcasting Service on September 22nd, but most residents believed the storm would switch directions and head north. It did not change course and between 6:30 pm on September 22nd and 6:00 am on September 23, 1955, winds reached 130 mph.[10] Because of the intensity of this hurricane, the name was officially "retired by agreement of the World Meteorological Organization," according to the *NOAA Backgrounder*.[11]

Other tragedies of the storm include the only hunter plane which was ever lost during a reconnaissance mission: "On September 26, the pilot lost radio contact at 10:15 pm, before flying into the then Category 4 storm. By 11:00 pm the US Navy classified the plane as overdue and the crew was officially reported missing. Neither the remains of the plane nor any signs of the crew have ever been found."[12]

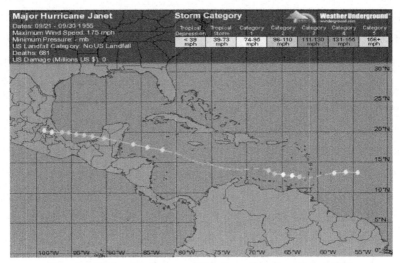
Weather Underground chart of Hurricane Janet's path in 1955

Lily's house was completely 'mashed to the ground' as the residents say and she was seriously hurt when she was picked up by the storm and thrown into a cistern just outside her home. She hit her head, suffered deep cuts, and a broken hip. She was bedridden for a long time as it was not deemed safe to transport her for medical attention to another island. She required daily visitation by a nurse to change her head bandages and to help with her personal care. These injuries would keep her an invalid for the rest of her life.

After the hurricane, Ellis built another house that he, Fedelin, and Lily lived in along with the children who Lily was raising. It was completed in 1956 using greenheart wood, which is the same wood used in many buildings on Petit St. Vincent, and glass windows, all purchased from Guyana, South America. Because Lily was hurt so badly, Fedelin assumed responsibility for the care of the children. Altogether, there were seven people residing in the home.

Because of Lily's accident, she could no longer take care of the island. She decided she was going to sell. Her son Ellis agreed this was the right course of action. Many people were interested in buying the island but the person who ultimately signed the agreement with Lily was Mr. H.W. Nichols, Jr.

CHAPTER THREE

THE NICHOLS

Mr. Harold Willis Nichols, Jr., and his wife Kate, sailed the Grenadines for many years beginning in the late 1950s. Willis, as he was called, was born in 1911. He graduated from Milton Academy, a boarding and day school in Milton Massachusetts and continued his education, graduating Harvard University in 1934. He and Kate married in 1936.

After college, he joined the Fox Paper Company in Cincinnati, Ohio until he began serving in the Army Air Corps as a first lieutenant. After his discharge as a colonel from the military in 1946, he returned to Fox Paper and presided as owner, president, and chairman of the board. Mr. Nichols operated this business until he sold it in 1971.

Kate and Willis Nichols 1942

Willis Nichols in Uniform-WWII

During the same time, he and his wife owned a Standardbred horse farm in Lexington, Kentucky where they raised and trained harness horses. Mr. Nichols continued to operate those two businesses for twenty-five years while also finding time to volunteer in community pursuits. Not only did he hold the titles of treasurer and national vice president for the National Heart Association for many years, he also served as president of the Cincinnati Opera House.

Being the great leader and kindhearted man he was, Willis also fulfilled the needs of local organizations, offering his time and effort as president and trustee of the Indian Hill and Madeira Volunteer Fire Department, chairman of the Ohio Heart Association, trustee and chairman of the Southwestern Ohio Chapter of the Ohio Heart Association, and chairman and trustee of a private primary and secondary school in Lexington, Kentucky. Until 1985, he also served as a director of the Hambletonian Society. Did this leave any time for leisurely pursuits? Sure! He loved to play golf and, of course, go sailing!

Katherine Harkness Edwards Nichols was born in 1910. She has an extremely full and exciting story as well. She operated a Standardbred nursery in Lexington, Kentucky for forty years called Walnut Hall Farm which was founded by her grandfather in 1892. In 1936, Kate became a world champion harness driver when she was named the leading woman driver of three 2:00 miles by driving the four-year-old mare Margaret Castleton to a 1:59 ¾ mark. Just a year later, the pair outdid themselves with a time of 1:59 ¼, with Kate being only twenty-seven years old! The mare remained an important foundation of their brood stock for many years to come.

Kate Nichols driving mare Margaret Castleton

Willis was an amateur harness racer himself. While enjoying the sport, both Nichols served as directors of the Little Brown Jug Society as well as the American Standardbred Breeders Association. Mr. Nichols held the role of director of the Lexington Trots Breeders Association for forty years and was also the board chairman of the same association. Other contributions to the sport included serving as board chairman of the Grand Circuit, president of the Horseman's Publishing Company for thirty years, trustee emeritus of the Trotting Horse Museum and Hall of Fame of the Trotter, as well as trustee of the Stable of Memories in Lexington. And the list goes on!

Katherine Harkness Edwards Nichols

Willis and Kate holding trophy when their horse Intruder *won the Hambletonian 1956*

Kate did not just have a love of horses; her endeavors included canines and cattle as well, raising and breeding Shetland sheepdogs and Charolais cattle. She held the title of president of the American Shetland Sheep Dog Association as well as serving as their secretary for fourteen years. She was also the founder and director of the Ohio Valley Charolais Association.

All of this while raising their four daughters [Martha, Kitty, Meg, and Beth] was quite a responsibility and accomplishment for any couple. However, this was not the end of their enterprising aspirations. Willis and Kate were just beginning an incredible journey that would enrich their family's already quite full life in tremendous ways. And this is where the history of Petit St. Vincent, as most guests know it today, actually begins!

Willis and Kate were avid sailors in the late 1950s and early 1960s. They would charter boats and sail among the Grenadine Islands as well as near Puerto Rico and the Virgin Islands. In February 1960, as they were sailing, they came upon Petit St. Vincent and noticed it for its extreme beauty. It was one hundred thirteen acres in size and was surrounded by almost two miles of white sand beaches. It had several hills offering magnificent views, "the water inside its lagoons was crystal clear . . . and the harbor was excellent. Further, it was uninhabited and privately owned."[1] The Nichols also liked the fact that the island was somewhat protected by the other neighboring islands of Petite Martinique, Union Island, Carriacou, and Palm Island.

They found the area to also offer a marvelous climate. The easterly trade winds blow across the Atlantic Ocean for three thousand miles from the closest coast in Africa and the water temperature only would vary within the year by about four degrees. Sunshine was prevalent most every day with rain only lasting for short periods except during a few months of the year where heavier showers could be expected. Nichols also stated in his written notes:

"Travel in the Grenadines in those days was not easy. There was a mail boat which touched a few of the larger islands once a week, or some of the island trading sloops or schooners would take deck passengers if you were going in the direction they were. Other than that, charter sailing yachts could be arranged for out of St. Thomas or Antiqua and a few were beginning to base themselves in Grenada. Shore accommodations ranged from non-existent to very primitive." [2]

Between the years of 1960 and 1964, the Nichols continued to sail the Grenadines, completing their tour of all the islands between Trinidad and Puerto Rico. "The people of the Windward Islands seemed to us to be outstanding. They are proud, honest, industrious, generally very fair in their dealings with others, and we found them very friendly . . . we found nothing which could compare with the Grenadines. They are the little gems of the Caribbean."[3]

In the spring of 1964, when they decided to return to the Grenadines, their usual boat was not available so they were furnished with a different charter; a 77' schooner by the name of *Jacinta*. It was owned by Douglas Terman, usually referred to as Doug, and Hazen Richardson, known as Haze. The Nichols family was the first charter *Jacinta* took out. Willis described it as "beautifully built and very roomy, in the old-fashioned style. Doug and Haze turned out to be delightful people."[4]

The Nichols returned in 1965 aboard *Jacinta* and inquired about purchasing the island as Petit St. Vincent was exactly what they had been looking for. According to Mr. Nichols, "It was an out island, had a good harbor, excellent beaches, and good hills which were not too high to climb, and which offered spectacular views."[5] Other amenities they found suitable to their specifications were a local population on the neighboring islands who were "intelligent, clean, extremely decent, and very friendly."⁶ They went ashore on the nearby island of Petite Martinique to inquire where to find Ms. Bethel, the apparent owner of Petit St. Vincent.

Doug Terman, Haze Richardson, Mr. and Mrs. Nichols, and Dave Corrigan

Kate Nichols with children from Petite Martinique 1966

Petite Martinique is a small island of five hundred eighty-six acres, then with a population of seven hundred and they noted that it was mostly inhabited by women and children with "some two hundred men [away at sea] in native trading vessels, tramp steamers or ocean liners or were working overseas and sending home money regularly."[7] According to the Nichols, "it was said to be the richest island per capita in the Caribbean." It was a far outpost of Grenada and they did not, therefore, get much attention from the government, nor did they want to. The sole representative of the government was Mr. Roberts, who was not only the schoolmaster but the postmaster as well. Without government intervention, the residents "lived by a strict moral code which they collectively enforced themselves."[8] Work on the island encompassed growing sea island cotton, *Gossypium barbadense*, and a few vegetables.[9]

A Grenadines legend, spoken often by many residents and known within the yachting community, was told by Willis in his notes, attesting to the independence of these great people of Petite Martinique:

In the days when these islands were British Colonies, much of government revenue was derived from custom duties. Human nature being what it was, there are always those who seek to avoid such taxes and as the wealth of the people of the ship building and ship owning island of Petite Martinique continued to increase, it crossed the mind of the colonial authorities that a customs officer should be stationed on Petite Martinique. It seemed that the local populous was not of the same mind. When the customs officer arrived, the entire population was gathered by a newly dug grave [dressed in their Sunday best]. "Who died," asked the customs man. "Nobody, yet," was the reply, "the grave is for you." The officer is said to have departed at once and whether or not the story is true, there is still not a customs officer on the island![10]

Willis, Kate, Haze, and Doug went to Ms. Lily Bethel's home to discuss the possibility of purchasing PSV. According to Mr. Nichols, their first encounter with Ms. Bethel began like this: "We were directed, and I might say, escorted, by about half a dozen children up a steep hill, which we negotiated by means of a very treacherous path, the kind where you sometimes go forward three feet and slide back two."[11] The group was very impressed once they reached Lily's home as they were warmly greeted by her sister. She asked that they wait patiently while 'she prepared' Ms. Bethel for visitors. As noted previously, she was bedridden from the injuries she sustained during Hurricane Janet. It is believed that Lily was in her seventies at the time. "She was very gracious" and, as Willis later wrote, "even if we hadn't done business with her with the island, we had met a very remarkable individual, and had made a new friend."[12] She truly upheld the title of "the undisputed Queen of Petite Martinique."[13]

Mr. Nichols, upon researching the ownership of the island, found that Lily's mother Mary, had indeed bequeathed it in full to Lily before her death, with the paperwork being drawn up by Mr. Tomms, Lily's lawyer in St. Vincent.[14] Since he was satisfied there was a clear title, he flew back to the United States for business and left Kate, Doug, and Haze in charge of the final details. They once again visited Ms. Bethel and although the Nichols considered the first asking price to be too high, they finally agreed upon US$50,000 as a suitable offer which Lily accepted.[15] The paperwork was signed on March 8, 1965 selling the island of Petit St. Vincent to H.W. Nichols Jr. Willis was fifty-four years old at the time and was just beginning this new, very large, undertaking.

And what an amazing journey it would turn out to be. The first step, according to the law, would be to obtain "permission to buy and hold land in the colony of St. Vincent"[16] since the Nichols were considered aliens of the country. This authorization had to come in two forms; one from the St. Vincent government and the other from the Foreign and Commonwealth Office in London. So the wait began and it is to Ms. Bethel's credit and unwavering character and honesty that she did not sell the island to someone else even though she had better offers during this time. According to Mr. Nichols, even though everyone was frustrated with the time involved in this process, Ms. Bethel told him that "she has agreed to a sale and would not go back on her word."[17] The Nichols felt this was an outstanding trait he had noticed of the people within the Grenadines and a true compliment to their character.

While waiting for proper bureaucratic approval, the family decided to purchase and send a wheelchair to Miss Lily. They did this out of sheer friendship because they considered her a "truly remarkable person,"[18] and that it had no connection with the fact they were in the middle of a business deal. Either way, it shows the extraordinary type of people the Nichols' were. Doug Terman was the person who delivered the modern, lightweight wheelchair which offered Lily the flexibility to venture once again outside of her home.

Other affairs needed tending to while awaiting the alien landholding license. As Willis' daughter Beth commented:

During this wait, they were by no means idle. Planning began – first thing they needed was water… so both my father, Doug, and Haze began researching solar stills. Next they made plans for a dock, and then a building. They still didn't have the license! [19]

Also, during this interval of time, Doug and Haze continued their chartering business.

Beth Nichols believes her parents initially envisioned the island as "a place for yachtsmen to get a meal, shower, ice, a drink or two, and possibly a room."[20] Around that time, yachtsmen found it hard to find places with such facilities. And she continues, "they thought it should be for people who wanted to get away . . .

not for people who had more money than God."²¹ However, it was made apparent by the Vincentian government that in order to obtain the alien landholding license, the property would need to be developed according to the Hotel Act and bring economic assistance to the area. So the decision was made at that time to develop the island as a resort.

Beth, Meg, Kitty, and Martha Nichols

CHAPTER FOUR

THE MAKING OF PETIT ST. VINCENT

At long last, in the fall of 1966, the Nichols were granted alien land holding rights to PSV by Sir James Mitchell, Minister of Trade, Agriculture, Labor and Tourism! Originally, the island was split in ownership between Willis Nichols as the cash investor, and Doug Terman and Hazen Richardson, who were to put up *Jacinta* to secure a share in the island. PSV LTD was organized in 1966 by Willis. Also involved was Harry Santen, who was the Nichols' lawyer until 1986. Through this corporation, stock was issued to obtain working capital for the island's development. "The capitalization of the corporation was composed of the island, improvements, and the schooner *Jacinta*."[1] The boat was now owned by the corporation, and Doug and Haze became the managers, being responsible for its maintenance and operation.

However, as plans were developing, Doug and Haze realized they would not be able to continue financially in this venture. "Haze and Doug were getting cold feet while they were waiting for the license and wanted to flip the island," Willis noted.[2] So the Nichols bought their share of the island for a small profit to Haze and Doug and the two became salaried employees of the island. The decision was made to have Doug remain on PSV to supervise construction and Haze would continue to sail *Jacinta*. At that time then, Haze and Doug were being paid to work on the island's development.

Haze and Doug began pulling together a construction crew to start building the first structure on the island; a storage building. It is the same building that is now called 'the dock house' and is still standing today. The local workmen were divided into groups and Haze and his workers began construction on the generator house. This was very far ahead of the times because none of the neighboring islands had, nor would have, electrical power for many years to come. At the same time, Doug and his crew completed work on the storage building which housed cement and tools, and then began the next project; building a jetty. While this construction was in the beginning stages, Doug and Haze lived on *Jacinta*.

It was noted by Mr. Nichols that he believed the water in the pond was not drinkable and therefore began devising a system to provide fresh water. Residents from Petite Martinique were still coming over to collect water to use in their homes since they were experiencing very dry weather that winter. Because the water was being used for people's needs, the concrete that was mixed that first season was prepared with sea water, rather than with water from the pond, as was originally intended.

During the time of early construction, one of the original workers, Noel Victory, otherwise fondly known as Goatie, was employed as a mason and plumber at the young age of nineteen. He tells the story of how he and the other crew members originally used the pond's stagnant water for drinking. They would clean out oil drums and each night before they finished work, would fill them with the pond's water. Ash from a fire of burned bramble was then poured on top of the water in the drum. By the time they started work the next morning, the cinders had settled to the bottom of the barrels and the water above it was clean and 'purified' for the workers use. They would skim off the top layers of water with a ladle-type scoop and would drink it without worry of contamination.

To contend with the situation of needing a consistent fresh water supply, the next project Doug undertook was the creation of fifteen solar stills to desalinize the sea water and make it suitable for drinking. In the book entitled *Adventures in the Trade Winds*, author Richard Dey explains how Doug had designed a "little masterpiece"[3] by purchasing a windmill in Grenada which drove a 12-volt alternator. This charged a battery, allowing a pump to run and bring in the water from the sea, while another 12-volt bilge pump blew air into the unit to keep the still inflated. Initially it was reported to have generated approximately one hundred gallons of fresh water each day and at its maximum, the production rose to about eighteen hundred gallons per day.[4]

Doug had never actually planned to stay with the island venture for long. He had told Mr. Nichols that, at best, he would stay on for two years. But by the spring of 1967, after only a year, he booked passage for his wife, child, and himself on a *Geest* banana boat and left the island.

At the same time, Haze had been visiting Mr. Nichols in Cincinnati developing plans and making various purchases for PSV. One historical acquisition was the boat named *Striker*. It is 34' long and considered a sport fishing boat that had two 108 hp Ford diesel engines. Haze checked it over in Chicago, had a survey and an inspection prepared, and then had it shipped by tractor trailer to Miami. After a few modifications were made per Mr. Nichols specifications, Haze and a Vincentian who was hired, named Joe Brown, flew to Miami where they started their long eighteen hundred mile journey to PSV. Joe Brown was actually one of the proprietors of Corea and Company, a large St. Vincent department store and import/export company. Amazingly *Striker* is still being used on the island today to transport workers to and from Petite Martinique.

Doug was in transit away from the island and Haze was returning to PSV when they met in St. Lucia. In one night, Doug tried to fill Haze in on all that had happened on the island while Haze was away, which encompassed about a three month period. When Haze finally reached Petit St. Vincent once again, he settled in as resident manager just as the major part of the construction was going to begin.

In early 1966, Willis and Kate Nichols came back to the island and were discussing plans for construction. One day, a man named Arne Hasselqvist, who was from Sweden, sailed to PSV on his 26' boat and asked Haze if he needed a builder for the island. With all the planning and clearing already going on, they surprisingly still did not have one in place. Arne was soon hired and would often sail from neighboring Palm Island, where he was planning to construct his home, to discuss these plans with the Nichols. Doug and Arne decided the volcanic rock from the island should be used in the construction. Because the dock house had recently been finished, and the look of the stone building was so attractive, it set the stage for the entire development of the island's subsequent buildings. With so many people involved, the Nichols decided it would be best to retain control of making all final design and plan decisions.

It is interesting to note that the original name of Palm Island was Prune Island. It was first developed by John and Mary Caldwell when they leased the island for ninety-nine years beginning in 1966. Under the terms of the lease, it was stipulated that for the payment of US$1, the island could remain in their possession if they constructed a hotel which would provide local employment.[5] The reason the name was changed to Palm Island was because of John. He was nicknamed 'Johnny Coconut' since he had been known to go around to various islands planting coconut palm trees. It is believed that Johnny Coconut planted some of the coconut palms that are on Petit St. Vincent today!

The main pavilion's distinctive design, according to Willis Nichols, belongs to Doug Terman. He was very excited about showing off its intended plan so he devised a model for everyone to see. He meticulously worked on it by patiently putting together small pieces of wood. When it was finally completed, its presentation did not go as planned. To everyone's unfortunate surprise, Doug's dog Sam, who apparently was the most excited of all, triumphantly sat on the whole thing; it was mashed to the ground! In spite of Sam's enthusiastic flattening of the model, construction went ahead including the addition of a twenty thousand gallon cistern [today used as a wine cellar] being added to hold water. The center of the pavilion radiated with beams made of purpleheart wood, which is considered to be one of the hardest woods in the world. It projected to eight points which were supported by stone walls on each end. They found that by having the stone walls, the pavilion could visually be divided into smaller areas, rather than it seeming "as though you were sitting in a barn."[6]

The initial 'honeymoon houses' as they were called, were designed entirely by Arne and ultimately ten cottages of this type display his creation. The six beach cottages on the island's north side were developed completely by Haze from a recollection of a home located in Westerhall Point, Grenada which he had always admired. The three duplex double cottages [containing six units] were created through a collaboration of Arne and Haze. In total, twenty-two cottages of three different designs were built.

View of the pavilion from the road to the dock house

"Mrs. Nichols and I rejected dozens of other sketches and suggestions, believing the architecture that emerged blended well into the landscape and thus presented one basic style of architecture and not a conglomeration," noted Mr. Nichols.[7] They decided to build the units, not as hotel rooms but rather as comfortable, large cottages strong enough to resist hurricane strength winds. In fact, all the buildings meet these specifications although by 1966, only one hurricane had previously come through the island, which was Hurricane Janet as noted in Chapter Two. The Nichols gave a tremendous amount of credit to Mr. Hasselqvist for conceiving and carrying out the plans. "Arne Hasselqvist . . . turned out to be a very efficient construction engineer," noted Willis.[8]

Arne went on to run an international architectural design firm where he created homes throughout the Caribbean but is best known for his custom homes in Mustique. He eventually designed and constructed his personal house on neighboring Palm Island, and "he is considered one of the best architects and custom homebuilders in all of the Caribbean," notes *Paradise News* in July 2000.[9] He has designed and built homes for people such as American fashion designer Tommy Hilfiger, musicians Mick Jagger and David Bowie, and Her Royal Highness Princess Margaret.

Arne began work in 1968 under Mustique Island's owner, Colin Tennant. He worked along with Oliver Messel, a British set designer, to convert an old warehouse that stored cotton into what is now called the Cotton House Hotel. It served as the concept for the entire island. He is credited for designing and constructing the Mustique Primary School, shops around the island, Basil's original bar, worker's bunk houses, the airport, and the roadway system throughout Mustique. His innovations brought work to countless Vincentians, ultimately employing a three hundred person maintenance and construction business. His designs always incorporated the use of local materials and as such, "his results are practical, cost effective, and stunningly attractive."[10] His clients know that Arne holds the highest regards for nature and the environment, and it shows in his work. Even though he earned a living from selling his designs and custom homes, he still worked closely with the St. Vincent government to make sure the number of homes on Mustique would be bound to no more than one hundred. In 2000, it was announced that Arne Hasselqvist entered a joint venture with Old Bahama Bay resort on Grand Bahama Island.[11]

A poem about Mr. Hasselqvist is written in a book called *A Glass Half Full* where author Felix Dennis speaks fondly of Arne. He also speaks of how "his sales technique towards prospective buyers was legendary, and inevitably contained the triumphant clincher: ". . . and it will all be finished by Christmas week! What Arne often did not specify was which Christmas he had in mind."[12] "It was the vision and work of Tennant, Hasselqvist, and Messel that created the blueprint that is Mustique today."[13] PSV is very lucky indeed to have had such a great architect design many of its buildings.

Even though the construction was going well, a real challenge continued to be communications. Staying in touch with each other to address the many questions and concerns that would arise was difficult. There was no phone communication. Ham radios or ship-to-shore radios were used as well as voice-recorded tapes or telex exchange to send messages back and forth to the different parties. Mail delivery to the United States would take two months which was not practical to use and if you did mail something, outgoing and incoming mail would have to be taken to Union Island. Therefore, Mr. Nichols would visit the island numerous times every year to make sure things were going as planned and the operation was running smoothly. Every detail was approved by Mr. or Mrs. Nichols before it could be implemented.

By 1967, it was decided that the island needed a manager and Bill Eberhart was chosen. Although he was retired at the time, his background included the construction and operation of a motel in Florida, and had assisted in the organization of a small hotel in Marigot Bay, St. Lucia while he had been sailing the Leeward Islands. Bill spent the fall and next spring with the Nichols in Cincinnati helping with the set up of PSV, such as purchasing, handling the books, and expediting work.

Another instrumental person in the initial stages of the island's creation was Howard Melvin, who at that time owned and operated the Heritage Restaurant in Cincinnati. Mr. Nichols became acquainted with Howard when he assisted him one year with chartering a yacht in the Grenadines. Working with Meg Nichols, they travelled throughout the United States in search of suitable equipment for the resort's kitchen. All pieces needed to be compatible with the climate of the Caribbean which can be extremely harsh on metals. It was also necessary to find proper bed furnishings, suitable as well to the environmental conditions. The Stearns and Foster Company actually designed special mattresses for the cottages which were resistant to mildew and rust. Both Willis and Kate were very happy with their final choices.

During this same winter of 1967, the Nichols did an extraordinary amount of shopping in New York and a few other states choosing what type of furnishings and draperies would adorn the cottages, which china, linen, and glassware would be used in the dining room, etc. The coordination of accumulating every item and arranging transport was undertaken by Bill Eberhart and it was noted by the family that he did a fantastic job.

However, there is something to be said about the phrase 'the best laid plans.' When the cooking equipment arrived on the island, which was accompanied by specific installation floor plans, none of it fit! At that time, communication about the kitchen project was in the form of cables, so the conversation went back and forth trying to find out how this mistake could possibly have happened. It was finally determined that the plans which had been sent to the island with the equipment were not the exact ones Arne had drawn up. Because a blueprint machine, which was necessary for printing the plans Arne had made, was not available in Cincinnati at the time, the Nichols drafted what they thought were identical plans "based on our understanding of what was going on at Petit St. Vincent."[14] But the floor plans didn't match!

> "The discrepancy lay in the fact that we had figured 30, 35, and 40 foot beams projecting from the rim of a 20 foot cistern in the center of the building, whereas, as actually constructed, it was a building with these beams projecting from the center of the 20 foot cistern, which resulted in the building being 10 feet smaller on the radius that had been planned, or at least intended by us. Arne explained that this had become necessary due to tying the roof of the pavilion down against 135-mile winds, but somehow nobody bothered to tell us in Cincinnati."[15]

View of the pavilion 1968

View of the pavilion from the boat dock 1968

Despite that confusion, the pavilion was able to open February 1, 1968 with the bar beginning official operation and food service following on March 1st! The magnificent bar in the pavilion was created by Goatie's brother, Percy Victory and another worker named Lee Providence. It was nice for yachtsmen because Petit St. Vincent and Palm Island were now places where they could buy a drink and, as Haze always boasted, they could buy a loaf of bread and take a shower! Also, by this time, the hilltop cottages #1, #2, and #3 were finished. Yachtsmen had the availability to rent two of these cottages for overnight accommodations and Arne, his wife Anita, and baby Lukas occupied cottage #1 [renamed later to cottage #20] instead of traveling back and forth to Palm Island each afternoon as he had previously been doing.

A
SPECIAL WELCOME
TO YOU,
OUR FIRST DINNER GUESTS
AT PSV !

DINNER MENU

Oxtail Soup

Lettuce & Tomato Salad

Filet of Steak with Onions
French Fried Potatoes
Carrots   Christophene
Fruit Salad and Cream
Cheese and Crackers

Coffee   Tea

February 29, 1968

First menu offered at PSV

When the pavilion was formally opened, Haze threw a party for the two hundred thirty men working construction as well as the entire yachting fleet which could be contacted by radio! It has gone down in history as "one of the greatest celebrations ever held in the Grenadines," according to the Nichols.[16] Named the 'Sheep and Bull' party, the rum was flowing, ox and sheep were roasted and this party was talked about for many a year. Even though Mr. Nichols was not present, the stories received by him from the staff and the people in attendance showed that it was 'quite a party.' The rum was served from a large iron sugar kettle which was on the island when it was acquired. Yachtsmen called it 'people pot' probably because it was large enough to hold something the size of a person. Later it was used in many cookouts and could make enough food for literally hundreds. This was the only item that accompanied the island's purchase and for many years it sat at the foot of the steps to the pavilion.

Another new addition that spring season was a steel band. The construction workers organized themselves and named their band *The Island Waves*. They competed once at Carnival in St. Vincent with a number of other bands and returned with a first prize victory!

By the end of 1967, most of the major construction was finished and it was decided that only seventy-five of the original two hundred thirty crew members would be kept to work on completing the cottages. Originally, many had been brought from St. Vincent as skilled laborers such as masons, plumbers, and electricians. These men had been living on Petite Martinique or Union Island and would be transported to work daily by boat. The men that were leaving were paid their final wages and a boat named the *Federal Queen* arrived to take them home. It travelled on to Union Island where it gathered more workers and then set off to St. Vincent.

It was a couple of days before Christmas and the workers were anxious to get home to enjoy holiday celebrations. However, tragedy struck. The *Federal Queen* had adequate capacity to carry approximately thirty but as more and more people boarded, the numbers grew to over one hundred thirty some estimate. On the journey to St. Vincent, the boat capsized during the night, killing all but thirty-eight who managed to swim to shore. Many of the fine men who were responsible for the creation of the cottages and other fine features as they now stand perished that night. Goatie's life was spared when Haze asked him to stay behind and not get on the boat. Unfortunately, Goatie's brother, James Victory, was one of the men who lost their lives. It was a horrible tragedy that will never be forgotten and each man's contribution to the island will always be valued and remembered.

CONCEPT OF THE RESORT

Willis and Kate Nichols decided that Petit St. Vincent should be created for the type of individuals that would like to escape the stress and hassle of telephones and televisions and therefore decided to offer it as a "quiet, family-type resort"[17] where everyone could enjoy a peaceful vacation. PSV offered activities such as sunning, swimming, snorkeling, fishing, day sails to neighboring islands, and water skiing. Also available if one chose were dinghies for sailing and glass bottom boats for those 'not-so-thrilled-about-snorkeling' folks.

Although Mr. Nichols loved playing golf, it was obvious the island's size could not accommodate a golf course. He decided though to have a tennis court built to amuse guests and offered other popular games such as shuffleboard, ping pong, darts, volleyball and badminton, croquet, and horseshoes.

The Nichols wanted to serve the best foods possible since fine cuisine was considered an important part of PSV. It was decided to import fresh meats from Miami with the fish coming from local sources or from St. Vincent. One of the local fishing vessels was named *Marguita* and rather than fishing by rod and reel, handlines were thrown. One can see this practice called handlining still in use today. It is considered one of the oldest forms of fishing.

It was difficult to obtain some fruits at all, such as oranges and limes, and many others were only available seasonally. However, in order to offer variety for the guests, these foods were flown in by charter plane when needed. Fine chefs were chosen and brought in from other countries to create delicious meals. Breakfast could be delivered to the cottages, where many guests seemed to enjoy eating especially on their outdoor patio, or could be enjoyed in the dining room. Lunch consisted of a buffet in the pavilion or a picnic lunch could be packed for them to eat on the beach or to take with them on their day trips off the island. Dinner usually offered guests a three-course meal with the option of two entrees and was served in the pavilion.

The design of the cottages included the use of native volcanic rock and timbers made of purpleheart wood. They had a dressing room and bath with two wash basins. The bedroom measured 12' x 16' and included two queen-sized beds, two finely upholstered chairs and a very large table. There was plenty of room for relaxing. The living rooms had two convertible sofas and sliding glass doors lead to a large partly-covered patio offering both sun and shade. The cottages were not air-conditioned but were situated to take advantage of prevailing winds, which were especially strong in winter.

Their designs guaranteed that each cottage was built with ample space around to ensure guests' privacy. They are connected by a road system that guests can use to access other cottages or various areas on the island. Transportation was limited to a pair of Mini Mokes which are manufactured in Britain. The staff drove by every thirty minutes in these mokes to see if the guests were in need of anything. However, in order not to disturb guests, a privacy system needed to be designed.

The original arrangement was a bamboo post four-foot high which was placed outside of every cottage. A mailbox-type container painted red was hung from a rope attached to the post, with a cap at the end to hold its contents. Forms for room service could be put in the mailbox. Topped by the red cap, it signaled to the driver that the guests were in need of something.

The flag system, which was developed next and unique to PSV, can be credited jointly to the Nichols, Doug, and Haze as a way to communicate with guests without disturbing their privacy. It was necessary to devise a slightly different system as the vegetation was growing taller around the island and the mailbox was difficult to see at times. It was then a yellow flag was added and when it was raised indicated the room service form was ready to be picked up. That idea worked for the next couple years until it became apparent that many guests sometimes wanted total privacy. And what better way to signal their desire than with a flag! So red was the chosen color and now the total flag system as it is still used today was complete. When the request slip is filled out, put into the mailbox and the yellow flag is raised, it is time to pick up the order but when the red flag is raised, the driver knows the guests wish their privacy to be respected and they will not be disturbed.

Having the ability for the guests to leave home and reach the island on the same day was considered by the Nichols' to be their ultimate goal. The options in March 1968 were limited to flying into the newly opened airstrip on Palm Island. *Liat* was the name of the airlines that flew there with a small plane service offered between Grenada and St. Vincent.

One major struggle the airline had, however, was there were a total of twenty-seven islands to accommodate with their five large airplanes and a number of smaller planes. Their greatest limitation stemmed from the fact that airlines are not able to fly before sunrise or after sunset into the regions mostly unlit airports. And because of this, they could not always arrive in time to link up with jets arriving from Barbados.

There was an alternative plane service which flew between Barbados and Palm Island and could be chartered. It was called the New York Gateway for residents near New York City. The flights left between 8:00 am and 9:00 am to Barbados where guests could take the charter flight to Palm Island and be picked up by boat for transport to Petit St. Vincent in the afternoon the same day. The return trip was just in reverse, allowing the passengers to get back to their home the same day.

Others arriving from the U.S. had two choices, but neither one allowed them to get to the island the same day they were departing from home. One choice was to fly to New York one day ahead, then take the New York Getaway as described above. The other choice was to fly into Miami the day before, stay overnight and then fly into Barbados which entailed another overnight stay because the planes arrive there after dark.

But people figured out their flights and guests arrived without difficulty. Over the years, the flights have become more coordinated and guests are able to arrive much easier.

THE GRAND OPENING

The winter season of 1968-1969 was greatly anticipated by all and the cottages were waiting for their first guests to arrive! Everyone was looking forward to the opening of the resort and all cottages were booked for the Christmas and New Year season. However, there was something very important that was missing in some of the units. Something very important indeed . . . and that was plumbing!

Because the original plan of the Nichols was to build ten cottages on the island, and during the course of planning, the number rose to twenty-two cottages, not all of the plumbing had arrived. That year there was a dock strike on the East Coast of the United States and the plumbing which was scheduled to go out never made it. This meant that the supplies would have to arrive by air freight rather than by boat. The shipment was packaged to be carried as freight on the planes and so it left New York without a problem. Unfortunately, the Nichols found out that upon its arrival in Miami, all plumbing materials would need to be repackaged into smaller-sized crates for the next leg of its journey to Trinidad. From that point, a schooner would carry the supplies to the island.

The Nichols hired a crew of carpenters to take apart the shipment in Miami and put all the items into different crates. They waited for the plane to arrive but strangely, the entire load never showed up and no one could tell the bewildered group where the cargo was. "The shipment . . . simply disappeared off the face of the Earth,"[18] Mr. Nichols noted. An entire week went by with no trace of the plumbing before it was finally found in Atlanta, Georgia! Meanwhile, the cottages were still booked and there was no plumbing in them. Once the shipment finally arrived in Miami, everything was repackaged and sent on its way, reaching Trinidad ten days ahead of the guests' arrival. There was a schooner luckily available to <u>immediately</u> transport the materials to Petit St. Vincent, which is apparently a very rare occurrence in the West Indies. And as the guests walked into their cottages, the last of the plumbing was connected! The resort of PSV officially opened!

THE MAKING OF PETIT ST. VINCENT

First Mock-Up Brochure

Beach cottage taken in April 1968

Cottage #20 taken in April 1968

Rate Schedule 1968

ISLAND TREASURE

Kate and Willis Nichols enjoying their resort 1980

CHAPTER FIVE

ISLAND CONSTRUCTION

ISLAND CONSTRUCTION

How the resort became what it is today was a major undertaking. It must be acknowledged as a labor of love by the many men who worked tirelessly to make this the island many call 'home'. They had used limited resources to build an unbelievable world known as PSV.

The crew started with a man named Noel Victory. Those of you who visit the island know him affectionately as 'Goatie.' He was employed at the Central Housing and Planning Department of the St. Vincent government when Haze came to St. Vincent looking for workers. He hired Goatie, along with another man named Dennis Rose, to work on the island as masons and plumbers [mentioned previously in Chapter Four]. Dennis was one of the youngest tradesmen ever employed on PSV, beginning his work at age seventeen.

On October 15, 1966, Haze, Goatie, and Dennis boarded the mail boat from St. Vincent to Union Island with tools in hand. The trip left at 9:00 am and arrived in Union at 8:30 pm, as the boat had to drop mail at all the various islands along the route. A small boat named *Daisemon* was waiting to take them to Petit St. Vincent. It was skippered and owned by Inkleman Stewart. The four of them had to beach the boat since there was no dock on PSV. The first thing Goatie saw was a tent Haze had been using as an office which, he quickly found out, was going to be his home for awhile. He, Dennis, and Mr. Stewart stayed in the tent for the night after having their first meal with Haze of pork and beans, cooked on a little camp stove. After dinner, Haze took the dinghy and stayed on *Jacinta* overnight with his girlfriend. Breakfast the next morning consisted of crackers and salmon and was served up courtesy of Goatie and Mr. Stewart!

Haze returned the next morning and so did a boat from Petite Martinique with about thirty workers. They had been hired from St. Vincent and were staying on

PM where they would live for the remainder of the time they were working on Petit St. Vincent. Some additional men lived on Union Island during construction and would arrive each day as well by boat. Goatie ended up moving to PM where he stayed at night, going back and forth each day. As Mr. Nichols remarked in his written notes;

> *"The construction period was about to begin. Both Doug and Haze possessed good engineering backgrounds and their skills nicely complimented each other. They were confident that they could handle the construction of the project themselves."* [1]

The first of Doug's projects, as mentioned in Chapter Four, was to build the dock house, as well as another very important assignment; the creation of solar stills to produce fresh water. Haze's initial projects were building a dock, a generator house, a tool shelter, and a manager's house.

Dock house with manager's house above in the distance

The dock house was used to store the cement and tools to keep them dry. Doug and Goatie, who was a key part of his crew, built it to completion one month later. They used pitch pine from St. Vincent, which is a variety of pine with three leaves that produces pitch [resin] and is native to the Eastern Seaboard of North America. They had a hard time getting the material but eventually did through Corea and Company.

The only tools the men had were the ones they brought with them or the ones Haze supplied. They were mainly hand tools such as stone hammers, saws, cutlass' [similar to machetes], a wheelbarrow, some twine 'to pick a straight', and small pieces of steel to clean out between the stones. The only other piece of equipment the men had at their disposal was a second hand Land Rover which was ordered from Grenada and delivered on a schooner. Mr. Nichols tells of how the vehicle "was lowered by *Jacinta*'s boom to planks strapped to a dinghy which was then pulled as close to shore as possible. Our newly acquired labor force then attached ropes to the Rover and pulled it into the water and up on the beach."[2] This Rover was the only form of transportation available and was useful to the men to move things around the island. The men and Doug decided to name it She Vex.

In order to build one of the most important features on the island, which was the dock, even more ingenuity had to be used. There were no pile drivers which could be found in the area. However, Haze is credited with devising an idea to construct the piles without use of authentic equipment. Being the extremely creative thinker he was, he developed a plan to pour a triangular concrete footing which sat around a hollow pipe. The footing was then placed by hand vertically with a hose being connected to the very top of the pipe. The hose shot a very high pressure water stream which allowed the pointed footing to very quickly embed itself into the ocean's floor at pretty much any depth they decided was necessary. Without this resourceful man, who knows when there would have been a working boat dock?

One day during the dock's construction, a potential life altering event happened. Haze and Mr. Daman Frederick [also known as Dumbar] were working underwater and a piece of Haze's clothing became entrapped in the machinery; Haze could not free himself. Dumbar jumped in and somehow got the clothing untangled and was able to help Haze to the surface. Without his quick reflexes and thinking, Haze might have been lost that day. Mr. Frederick is a hero for sure!

In order to continue with the construction, Haze needed even more workers. He would tack a notice on bulletin boards in St. Vincent saying there was work available if anyone was interested. He mostly hired the tradesmen from St. Vincent and unskilled laborers from Petite Martinique. The residents of PM were from fishing and boat-building families and therefore usually did not have the trades' skills the men on St. Vincent had. Instead, they made up the important labor crews that would be an integral part of the island's creation. The entire team soon totaled close to two hundred forty men. And to get them all to the island was another amazing feat. Every morning three or four boats would pick up the workers, sometimes as early as 6:00 am, from where they were staying in Union or on Petite Martinique and arrive at PSV around 7:30 am. They would work until about 5:00 pm and then be taken back home by boat. That made for an incredibly long, hard day.

Workers arriving at PSV 1968

Workers arriving to work [taken from Jacinta*] 1968*

The second structure to be built was the generator house. Haze was in charge of that project. Goatie, Dennis, and Haze worked together with twins John and Ezekiel Decoteau, better known as JD and Knock It, along with many other crewmen. It took three long months to complete and is made of the island's volcanic stones. The timbers used for this building and subsequent others were ones they purchased in Guyana.

Constructing the walls was an enormous job itself and required a great deal of skill. But how did the workers get the stones in the first place? They had to be gathered by hand! Goatie chuckled, "Stone was like rain here."[3] Many men walked around picking up loads of individual loose stones but getting larger stones was a much bigger problem. Since there was no blasting used to unearth them from the ground, they had to be individually dug up with pick axes and shovels. They made a 'quarry' on the west end of the island, which can still be seen today. Trees were taken down to open up an area where this 'excavation' could begin. During this construction phase, many artifacts were found near where the manager's house was built including pieces of pottery left long ago by past inhabitants of the island.

The stones would then be taken to the job site. Each one was then carefully chosen for its proper placement on the walls of the cottages and other structures. It was picked up, held, and spun in their hands to see if it was suitable for use. Almost every rock had to be hit with a hammer and broken into the proper size. Next, the stones were secured with mortar which was a mixture of sand from the beach and either salt or fresh water. The wood from the manchineel tree was sometimes also used. It was burned to ash and then mixed with conch shells to sometimes make cement. The white lime it produced was so strong it would burn the hands if touched. So although it was very effective for its intended use, it was necessary to take much precaution when handling. The fumes of the manchineel could be very toxic as well.

Building each wall required technical skills from the masons. Goatie recalls how each wall of each building was constructed to be one foot thick. Two men would work facing each other, one on each side of the wall. They each had six inches in front of them to work. One man would set a stone, then the other, alternating as they went up. Goatie explained it was important to work with someone you liked otherwise you could "easily mess the other guy up"[4] by setting the wrong-size stone which would push his side out or in!

There was no heavy machinery or power tools available to these men. And the only blasting that ever occurred was when coral was removed to provide beach access. The crew used hand tools to clear the land for the cottages. Anything that was not needed such as brush and bushes was burned. To build the roads and to make cement, the rocks not suitable for structures were smashed into gravel using just raw strength and hammers! With the technology and power equipment available today, it seems almost an impossibility these jobs were actually completed.

The wages that each man was paid in 1968 for a day's work depended on whether he was considered a laborer or tradesmen. However, most of the men were laborers earning anywhere between EC$3.00 and EC$6.00 each day. The higher paid workers earned between EC$10.00 and EC$11.00 each day. There were only a few men earning the highest rate.

The necessity of acquiring clean water to drink was extremely important. As previously mentioned, the pond contained stagnant water which, in order to be cleaned, had to be filtered through ash in barrels. Doug Terman is credited, as you recall, for designing the solar stills that produced fresh drinking water for the island. This design used a pump system to drive salt water into a plastic dome-type structure. It was placed in an area considered the hottest spot on the island which is located on the southwest side of PSV, just off the beach. It had 12-volt battery blowers powered by wind generators which lifted plastic sheeting up to form a dome shape. Within the dome, from the intensity of the sun's rays, the water would vaporize and collect on the inside crown of the dome. The vapor, which was now salt free, would run down along the edges of the dome into collection tanks. The system was connected to a clock system on the pipes that was switched off each night.

Soon after, the manager's house was constructed, which freed up *Jacinta* for chartering, and "so the infrastructure was in place – now on to build a resort hotel,"[5] says Nichols! A mechanics shop and cement shed was each finished and were used as a location to make various repairs as well as for storage. About the same time, the cottages, the laundry room, and the commissary building were all starting to be built as well. The laundry room actually had two uses; it was also designed as a hurricane shelter if the need arose. The commissary building was where workers would obtain their food. They would make a small fire to cook their meals and Arthur Williams, better known on the island as 'Crab,' sometimes still cooks like this today!

View of manager's house from cottages 21, 22, and 23

In all, there were twenty-two cottages built, numerous roads, stone walls, and stone paths. There are storage buildings, workman's quarters, and manager's homes, the pavilion, water storage tanks, the kitchen, and the beach areas. All of these exist because of one man's dream which became a reality. And the reality was attained through the efforts, talents, skills, and perseverance of the Nichols family and all of those who served under their leadership. It was also noted by Mr. Nichols:

"I was thankful to the various Caribbean innkeepers; learning from them much that we should not do as well as many helpful hints on what to do. The challenge of

engineering reliable and trouble free mechanical and electrical equipment for operations under the unique climatic, logistic, and maintenance conditions faced by PSV brought us much valuable technical advice and assistance from many highly qualified friends in the U.S."[v]

These men worked every day to create a paradise for future guests to enjoy all while putting a tremendous toll on their bodies. That was a dedication which would not easily be found today but serves as an example of the type of people the West Indians are, as well as all the other nationalities that were represented while building the island. The workmanship they left behind stands as a tribute to the men. They produced the unbelievable quality that Petit St. Vincent is known for and what exists today is a mark of respect for their diligent efforts.

Kate Nichols at the PSV pavilion in 1968

CHAPTER SIX

THE BACKBONE OF PSV
HAZEN RICHARDSON & DOUGLAS TERMAN

Over the forty-eight years the island has been a resort, there have been a couple of key players whose individual contributions were essential in making the island what it is today. Three men stand out among the rest in the history of PSV – Willis Nichols, Hazen Richardson, and Douglas Terman. These three individuals, each in their own way, contributed tremendous amounts of time, effort, and skill as well as just plain love into the making of this resort.

But where did they come from? Who were they before they became the backbone of Petit St. Vincent? Mr. Nichols background was discussed in a previous chapter in great detail and he, along with his wife, were instrumental as the founders of the resort and were the motivation for all to follow in their footsteps. Doug Terman, although only associated with PSV for a relatively short period of time, was influential in crucial areas of PSV's creation as well. A little bit about Haze had been mentioned in an earlier chapter but exploring his beginning in life could possibly shed a clue as to how he was able to accomplish the tasks, and to achieve the amazing successes, that he did.

DOUGLAS TERMAN

Douglas Terman was a United States Air Force pilot from 1955 through 1963. He worked in Intelligence and flew with the night jet inceptor squadron. In the early 1960s, he became a launch crew commander of the ICBM Atlas missile [intercontinental ballistic missile]. He and Haze were very good friends and they finished their tour of duty at about the same time. Soon after, Doug went to find

Haze in Massachusetts discuss joining him in a business venture of buying a sailboat and beginning a chartering business. "Doug was an accomplished sailor and had owned several boats in the past,"[1] according to his wife Seddon Johnson, whom he married in 1980. "They were going to be charter skippers, somewhat influenced by a very popular TV series called *Adventures in Paradise* starring Gardner McKay."[2] So Doug suggested the idea they should go in together to buy a sailboat and run charters. And that is exactly what they did. Haze and Doug became captain and co-owners of sailing yacht *Jacinta* which they bought in New York. This yacht was built in 1931 and designed by Samuel Crocker.

Doug Terman 1989

Seddon shared an interesting tale that had been rumored about *Jacinta*. Apparently she "was haunted by a Spanish lady who could be heard occasionally roaming the decks when the crew was below," Seddon revealed. "When she was also seen by one of the Vincentian crew sitting in the stern . . . he jumped overboard!"[3]

Although Doug only stayed on the island until 1967, he became rather famous for his desalination design and ability to provide the resort with fresh drinking water. His work in the military probably set the stage for his ability to devise such a process without any formal training. "Although Doug always maintained that the best education he ever received was in the Air Force, he had studied some engineering and architecture at Cornell University," explained Seddon.[4]

After leaving PSV, he purchased a yacht in England named *Encantada*, a 50' sloop. "He motored inland from Calais to the Mediterranean and sailed across the Atlantic to continue chartering in the West Indies."[5] When his family decided to settle in St. Vincent, he started a type-setting company and also taught house construction to those less fortunate.

Encantada *with Doug standing in the cockpit*

Doug moved to Vermont in 1975 and became a well-known author. He started writing 'techno thriller' fiction stories as well as a book where you 'Choose Your Own Adventure.' A few of his well-recognized titles include *By Balloon to the Sahara*, *Shell Game*, *Enemy Territory*, and *First Strike*. In all, he wrote five novels and one children's book. To gain information for *Shell Game*, Doug met with and interviewed many people who had served in the Cuban Missile Crisis and *"First Strike* takes place, in part, on PSV," according to Seddon.[6]

Skipper of Jacinta *Tom 'the Walrus' Clarke mentioned in* Enemy Territory *[The hero's sidekick is called "Walrus" and is remarkably like Tom Clarke* [7]

Doug made many trips to visit Haze back on PSV and Haze had made many visits to Vermont, even purchasing land next to Doug there. "They were very dear friends, almost brothers."[8] Their friendship was able to continue although they were often far apart.

HAZEN RICHARDSON

Hazen David Kimball Richardson II was born in October, 1934 in Danvers, Massachusetts. He comes from a family that can be traced back many generations. Mr. David Richardson came to the United States in 1636 from England where they were farmers. They continued their farming operation and raised dairy cows once they arrived in Massachusetts. By 1900, Hazen K. Richardson [Haze's grandfather] was responsible for growing the business into what it is today. He expanded, adding many more cows, and with the help of his sons, Hazen M. and Ben Richardson, they started a dairy business in 1917. By 1934, the enterprise expanded to bottling and selling milk from over one thousand cows. "The Richardson's have milked cows in Middleton every day since his arrival [more than 300 years or 109,500 days straight.]"[9]

Haze contributed to the farm throughout his childhood. Every morning before school, he would get up early to milk cows. And every day after school, the same job was required of him. According to his family, he was a very, very hard worker even as a child. The farm has been run by a family member since it started in 1695 and in 1952 it became even more famous when Ben and Hazen M., Haze's father, began selling ice cream.[10] The Haze known by so many workers, guests, friends, and business associates is the 8th generation of this family, which is still in operation even today. His family remembers Haze growing up as a very bright, forward-thinking young man who was an adventurer.[11]

Richardson's Dairies

He attended the Phillips Andover Academy in Massachusetts for his high school education, graduating in 1952, and then went on to the Wharton Business School of the University of Pennsylvania. After attending college, Haze joined the

United States Air Force in 1957 and was a navigator and a B-47 bombardier. After leaving the military, Captain Richardson returned home to work with his father on the family's dairy farm. By the time Doug visited him, Haze had already realized the hard time he was having acclimating to a much quieter life after experiencing such an exhilarating time in the service.

Jacinta *in the harbor at Nelson Dockyard at English Harbour Antigua*

After Haze and Doug met the Nichols, even though they wanted to be involved financially in the making of PSV, they realized they did not have the capital it would take to see a project such as this to completion. So instead, they decided to work with the Nichols on the construction of the island. What Haze couldn't have realized at the time was how PSV would become his work, his life, and his legacy. After the resort opened, Mr. Nichols asked Haze to stay until a suitable manager could be found, but Haze was growing ever closer to the people of the islands and to the men who had worked so hard every day on the construction projects. Therefore he felt he needed to stay and take on the responsibility himself. He knew where everything was and how to get the best out of the staff. He was fair-minded and cared a lot for each one of them. He made sure they were well taken care of. If, for instance, a person needed a loan, they would usually receive it; if they needed money to buy a house, he would make sure their dream could become a reality.

Jennifer Richardson, Haze's wife who is mentioned in more detail in later chapters, remembered that both Haze and Doug had never originally planned to stay on the island and Doug ended up leaving after only a short time at PSV. She fondly remembers Doug as "movie star handsome." "They were both unbelievably gifted" and therefore an incredible team to be at the forefront of the resort's beginning.[12] Jennifer recalls Haze telling her that Doug had more knowledge about many details involved with the initial phases of construction and development of the island, but that Haze would spend nights studying civil engineering and the like by lamp light on board *Jacinta*. Haze had to take on an even more enormous task once Doug departed.

Haze came to this island which had limited resources. To assist him, he personally picked out most of the crew, going with his gut feeling, his skill in knowing what expertise to look for, and the recommendations of other team members. But as extensive as his military training was, it could not have prepared him for the inconceivable tasks at hand at PSV. So how did Haze manage to create the resort it is today? Jennifer says he put himself into the middle of it, got a hold of any book or any information that he could to get the job done, and began working. "If he didn't know how to do something, he learned how to do it. He was a very pragmatic, practical, and smart man." explained Jennifer.[13]

The workers knew him for the many roles he played such as manager, co-worker, and friend. In the island's newsletter in 2008 for the 40th anniversary of the island, it was written that over the years Haze had been an architect, laborer, boat captain, self-appointed coast guard, teacher, and builder. Plumber, carpenter, mentor, gardener, security officer, mason, electrician, mechanic, and navigator were also part of his duties. Exterminator, landscaper, entertainer, pilot, accountant, and even 'smuggler' were listed as part of his claim to fame! The last one shall remain understandable only to those who knew him well apparently! That is an amazing amount of hats to have to change into daily.

Because he was an avid pilot, he continued his love of flying by owning his own planes. The first plane he had for five years and was a NA Piper Apache aircraft. He kept it on a grass airstrip on Palm Island until Union Island's new airport was constructed. In 1979, he decided to buy a Piper built Seneca I airplane which he found in Massachusetts. He loved it so much he kept it for twenty years. Haze wrote in 1999 that he had just recently purchased a more comfortable airplane that was faster, had the newest avionic and radar equipment, and could carry much more fuel to sustain him on longer journeys.[14]

Haze and the staff members became very close over the years, with many seeing him as a father figure. What would cause people to hold him in such high regards? Haze was a man of very high integrity and he felt that each person deserved his attention and understanding. He encouraged them every day, wanted each person to know he believed in them, and that they were capable of accomplishing any task put in front of them or which lied ahead. He never made a person feel as if they were unintelligent if they did not understand something. He took the time to explain in detail what they needed to do so they could fully be aware of what was being asked of them. The staff remembers how Haze would sit around the bar with the workers, as one of them, enjoying each other's company and stories. They appreciated this personal approach he gave to the island and all of its workers.

The staff also knew him as a coworker because they labored together as a team throughout the years. They respected him because he always took the time to listen to what they each had to say, good or bad, to really understand each person and their needs, and then he would respond accordingly. Although he was their friend, he always requested that each staff member refer to him as Mr. Richardson, as a sign of respect, and the ones working on the island today still do just that. He considered many to be his sons, like Otnel and Goatie, Chester and Reynolds. And to them, he was their mentor, friend, confidant, and father figure.

It was not always just hard work . . . frequently there was time for relaxation. Haze and the staff would have fun together. "Haze would play backgammon with a post if it would play back with him. He played with me, he played marathon games with Doug Terman, he played with Slick, the chef; he would play with anyone. He was passionate about the game. He was a very, very good player, and it would drive him to the point of madness if I ever beat him, which I did, from time to time," Jennifer remembers with a chuckle.[15]

A reputation soon began to follow Haze when others heard about this incredible project he was able to achieve, and he could have easily left this so called 'fame' get to him. But instead Jennifer says he was the most modest man she had ever met. "He didn't think he was anybody special," she remarked.[16] And when Walter Cronkite wanted to speak with him, he couldn't imagine why! People would have to ask a lot of questions to get any information out of him. But it wasn't because he was anti-social; he was just a shy, unpretentious man whom Jennifer would have to drag out to events sometimes! His daily attire consisted each day of a polo shirt with its collar up and a pair of Bermuda shorts. Unassuming dress for an unassuming man!

Haze was also the most cool-headed person alive, Jennifer thought. "Cool under fire," he was.[17] She remembers so many times she was thankful for this gift of his, like when the United States invaded Grenada in 1983. Bombs were going off in Union, troops were trying to take over Grenada, and Haze just calmly told her that it might be a good idea to come down from the house and join him in the office. When she arrived, he explained what was happening in a very matter-of-fact but composed way. Jennifer was not quite so cool so she decided her plan of action was going to be: "I wrote out a will and then put all of my jewelry in a plastic bag and hid it in the walk-in cooler among the steaks and lamb chops!"[18]

And another time of composure, even more pressing and personal to them, was when they were flying in their plane and ran out of fuel at three thousand feet! Jennifer was overjoyed that he could stay so calm. She just looked at him and knew he would figure out what to do. Although she was petrified, as most people would be when you are handed a piece of information such as that, she just sat back and let him take care of it ". . . just like he always took care of everything."[19] The staff was grateful as well for his calm approach to everything. Like the time Otnel 'was taking driving lessons at night' as Haze and Jennifer called it and had 'a little accident.' Unruffled, Haze just explained that he would fix it with no big deal being made about it.

Because Haze was originally part of the purchase of the island, even though his role had soon reverted to manager of Petit St. Vincent, Mr. Nichols had put into his ownership documents a contractual agreement where Haze and Doug could acquire majority interest in the island. This gave them the ability to purchase the island first if something were to happen to him. After his passing, Haze found a financial backer in a long time guest of the island. Haze also spoke with Doug Terman who declined wanting any financial involvement with PSV.

"The story of how we found our investors to help us buy the island is priceless," Jennifer recalls. "Haze and I were such out of touch, naive people living in another universe, i.e. on our little 'rock' in the middle of the ocean, we didn't have the business

acumen to put together an investment group . . . what we did do, was so simple, so utterly unsophisticated and yet real at the same time. The first people we asked said, 'Sure, we'll give you the money.' It was all over in less than an hour and was a handshake deal at that point. They believed in Haze, and they believed in us as a team . . . we never let them down.'[20]

Jennifer is still very close to them even today.

Haze Richardson is considered by many to be the foremost contributor to the island's recent history. He took on roles he was unfamiliar with and had no formal training prior to beginning many of these projects. He accomplished unbelievable tasks. He was a great role model for the staff and will always exist in their minds as a fine man to look up to and a friend they were honored to have had. He was truly responsible for making one man's [Mr. Nichols] dream of owning an island a reality.

Hazen David Kimball Richardson II

CHAPTER SEVEN

THE MEN OF PSV

In addition to the enormous contribution of Willis Nichols, Haze Richardson, and Doug Terman who have been written about in previous chapters, there are many men who will go unnamed in this book. Because of their dedication to the island and painstaking effort they each have put forward into making the resort a reality, this book is being written today. Without them, PSV as we know it would not exist and therefore they are appreciated beyond words. They are a tribute to the integrity, ingenuity, and intensity with which they labored for so many years. To each man that has left their mark on the island . . . you are truly valued.

There are, however, two additional men who have contributed uncountable hours of their lives to the enrichment of Petit St. Vincent. They have been with the island since the early years and an extra, special thank you must go out to them. One started at the very beginning of the resort's development and the other soon followed. They have been loved for so many years by guests and continue to shine a ray of sunlight on the island today. Without these two men, the island would never be the same.

NOEL VICTORY

Noel Victory, better known as Goatie, began in 1966 as was previously mentioned in Chapter Four and Five. Haze had handpicked him and Dennis Rose and brought them to the island with him during those first glorious days of construction. Goatie has worked on the island ever since and is still going strong today. He is credited with being the longest-working man at PSV! He has had his hand on every building that has ever been built. He worked side by side with Mr. Terman and Mr. Richardson on every single project.

He speaks of how the workers carried everything by hand and labored unwearyingly each day breaking stone and moving it. Goatie would work with Dennis building stone walls and explained how the job was actually done. He would stand on one side, Dennis on the other, beginning first with one stone then the other. As Goatie pointed out in an earlier chapter, each man had to be polite to his partner because if he put the wrong stone on his own side, it would mess up the course for the other guy! Goatie thinks Dennis was a great man to work with as they were very like-minded and well-suited for the jobs they were doing. Therefore, he never caused anything like that to happen to Dennis . . . maybe another guy . . . but never Dennis!

Many workers on the island were related, so in the master time book in 1967 and 1968, there are numerous entries with the same surname. There were seven Adam's, five Alexander's, five Charles', three Bethel's, four Duncan's, seven John's, five Joseph's, four Providence's, five Robert's, three Victory's, and five William's. Wow! Talent certainly was extensive in those families; and isn't that fortunate for the island? But Goatie tells a great story about the biggest family to ever work on Petit St. Vincent. The surname is DaSilva and there were nine of them in the family working over the years; the boat captain one year was Michael, the second chef was Ezekiel, the mechanic is Lawrence, his sons are Larry, Kevon, and Gary, and there was Darwin, Elizabeth, and her sister's husband who also worked on the island.

It is a tradition in the West Indies to give nicknames to each person and not use their real names every day. So everyone in the master time book actually had a nickname they used on the island. Goatie explained how Ezekiel DaSilva's nickname 'Just Come' was given to him: The morning he arrived, a group of guys greeted him and asked him, "Who are you?" and he answered, "I just come." He was simply referring to the fact that he had just arrived, mistakenly hearing their question. But the name stuck and he was forever branded Just Come. Upon recollection of this story, Goatie and Otnel didn't stop laughing for ten minutes! John Corbett, manager during Haze's absences, says he and Haze actually saw that name, Just Come, on Ezekiel's passport years later!

Goatie liked the way work was done back then. Even though it was difficult, their efforts were very respected by the team leader. In the very beginning of construction, there would be about ten leaders who would each choose at least six men to work with. They were selected by their ability to complete their projects, make decisions, and think on their own. Goatie believes the reason the entire resort was able to be built in short two years rested on the fact that Mr. Richardson would ask a man to give an estimate as to how long it would take to complete a job. If the man said it would take five days to build a wall for example, and Haze accepted that time schedule, the project would move forward. And if it took the man only two days, he was still paid for five. In that way, the work was done more rapidly and everyone benefited.

Every year during the off-season, the workers were ready with plans in hand for what was going to be accomplished over the next few months. Haze had sat down earlier in the year with Otnel, Goatie, and other staff members, devising a strategy, having supplies ordered, and making sure everything was in place so the work could begin on time when the season came to a close.

After a long days work, Goatie recalls that Mr. Richardson would come to the bar, have a cigarette and a martini, and tell stories. He said he didn't believe Haze ever repeated one . . . each one was different, interesting, and often times funny. He speaks fondly of Mr. Richardson saying, "He was a good man; you could sit down and talk to him."[1] Goatie also thought he was very good with figures.

The bar was always hopping in those days and Goatie remembers how 'the blenders never stopped,' to the point of annoyance, when finally someone would stick a box over the top to quiet the racket! He remembers the first wine cellar was actually under the pavilion and in later years would be relocated to a new elaborate one by the restaurant. [More about the cellar in a later chapter.] There were Wednesday night Jump Up parties which were so much fun and something to look forward to. There are many stories associated with those parties which will remain as great memories to those in attendance, but a smile lights up Goatie's face when he recalls those famous Wednesday nights!

The regattas were other times of fun and excitement for the island, Goatie added. 'Clams' were the source of money during the weekend. The food was plentiful and the alcohol was flowing he recalls. The barbecue pit used during that weekend no longer exists but was located off the beach near the dock house.

Goatie built many more barbecue pits including the one by the reception area and the last one built near the Mr. Green pavilion. This fire pit was used for serving lunch to add extra variety to the menu. They grilled hamburgers, hot dogs, steak, chicken, and blackened fish and even made pizza. He also had constructed another fire pit which was down the road past the current beach restaurant. He says Mr. Richardson did not ask him to build it but the guests mentioned they thought the one being using at the time was too small so he made up his mind to fix that situation. Each night after an already exhausting day, he went down and built the barbecue . . . by himself. He said of all the beautiful structures he created over the years, this was his favorite piece of work and offered a sense of great pride.

Barbecue built by Goatie by reception area

Another one of his creations at the Mr. Green pavilion

As mentioned in an earlier chapter, a tragedy hit Goatie's family during the Christmas boating accident. His brother, James Victory, was one of the passengers on board that terrible night when the *Federal Queen* sank. He had just come to the island on holiday and had only worked for one month. Goatie didn't find out until the next day because there was no telephone and not many notifications by radio. Many of the bodies washed up on the island of Carriacou. A few lucky men who had been working at PSV made it safely to shore such as Edwin Providence and Arnold Duncan. The others will forever be missed but their work and dedication will never be forgotten.

Goatie is a man who cannot adequately be described with words. He truly has dedicated his entire adult life to PSV and continues to do so to this day. The new beach bar built in 2011 was named after him as a sign of respect, honor, and appreciation for all the years of service he has given. He said, "I loved the island so much that I wanted to do things for it, like it was my own home."[2] And it truly is. He built the walls and buildings and barbecues because of his devotion to Mr. Richardson and Petit St. Vincent. Who knows what would have happened if he wasn't chosen to work with Haze but thank goodness we never have to find out.

OTNEL SAMUEL

Otnel Samuel is another instrumental man of PSV who began his life on the island in August 1977, just ten years after Goatie and they have continued to work together ever since. At the time, he lived in St. Vincent in the Mesopotamia Valley where his family has a large farm growing bananas, nuts, fruits, and vegetables. A man named Harry DeGonville used to run a restaurant near the Samuel family's farm and would buy their produce. One year, Harry asked Otnel if he would like to join him for a holiday when he was going to be staying on Petit St. Vincent. Otnel gladly came with him and he has remained on the island all these years.

He started out as a groundsman and continued "because it fits and I just liked it so that is what I did."[3] Because Harry brought him to Petit St. Vincent, Otnel has carried the name "Diego" or "Diego boy" [pronounced Dēgo] ever since. He worked for two years as the groundskeeper and then he moved to storeroom duties in 1979. Within three to four years, he learned how to drive almost every piece of equipment they had then. He worked close to Mr. and Mrs. Richardson and whatever they needed they would ask Otnel to help. He became the person responsible for the inventory, but his job became more involved in many aspects of the island each year. He took each duty on as his own.

Otnel remembers how work used to be completed; every day everyone worked very long, hard hours but there weren't any complaints. He doesn't like the way he hears objections about the work today. He feels everything is much easier now and the workers of course do not realize how much harder it was then. All the work was done by hand and, as he points out, "Now they don't even have to go for water, they just open a faucet."[4]

Otnel has a very good background which is one of the reasons he believes he was hired. He is well-educated, speaks easily with guests, and is able to handle many different tasks at once. Otnel was promoted to Assistant Manager in 2010.

He was happy that Mr. Richardson had the forethought to teach him how everything was set up and how basically the island was run in case there were any problems in the future. Otnel could then order supplies, put the correct mark up on items for sale, and basically do whatever was needed.

Jennifer and Haze always called him Otnel, but sometimes substituted the name *Mr. Samuel* because it made him smile that "wonderful smile that travelled right up to his eyes."[5] When Jennifer saw that, she knew he was "really smiling on the inside as well as the outside."[6] Guests notice this quality as well and is part of what makes Otnel so special and charming.

Mr. Richardson was very organized Otnel thought. For instance, when it was time to purchase fish, they used to place that order two or three times per week. He wanted to use what was locally caught and in abundance that week so it could be offered for meals, just in different ways each day. Mr. Richardson used to analyze the details of what was available, especially during the lobster season, so he could make the most of what they would buy. Otnel speaks kindly and fondly of Mr. Richardson, who was almost like a father to him; he was extremely kind, very understanding, and very bright. He remembers once how Mr. Richardson helped assist with a boat accident: The vessel *Antilles* went aground near Mustique when she was coming from Barbados. He took over, got on the radio, and talked the captain out of the jam. To Otnel, it showed that Mr. Richardson was able to quickly make decisions and guide people on how to resolve troublesome issues.

During the years the barbecues were held on the beach, Otnel explained how the set-up was accomplished; plastic tables, chairs, and table settings had to be hauled down to that location as well as food and serving dishes, which took plenty of effort. However, there was one item which was a bit more cumbersome. It was the upright piano which was used for the evening's entertainment. Otnel said that each Saturday night, the piano was lifted up onto a jeep and driven down to the barbecue area! There were no trailers or equipment to pick it up or take it down off the vehicle, just brute strength! He said it took an hour and a half to set up for the entire evening and the same to clear out after. This was done every week for years. The piano entertainment stopped one distressing night when the upright was no more . . . it was accidentally dropped off the jeep during transport!

Throughout the many years Otnel has been working at PSV, many amusing things have happened. He starts laughing when he remembers a story about 'mountain dew'. Otnel would go out of his way to make sure all the guest's needs on the island were taken care of. On one particular occasion, he requested the services of Jeff Stevens, a private boat captain. One of the guests staying on the island at the time was very anxious to get his hands on some Mountain Dew soda. He loved it very much but PSV did not have any. Otnel knew the only place he could possibly get it was in Grenada. Jeff was sailing there with some guests so Otnel asked him if he could please pick up a case of Mountain Dew for a guest? Jeff said that would be no problem.

When Jeff arrived in Grenada, he hired a taxi and told the driver he needed to pick up a case of mountain dew and asked if he knew where he could get some? The taxi driver drove a long way, up winding paths around the island and finally stopped at this rather run down place. Jeff went to the door and a man answered. He told the man he needed to buy a case of mountain dew and asked him if he had any? The man went to the back and brought out a number of filled bottles. Jeff paid him and left.

Jeff was very happy he was able to help Otnel out and presented him with the case of 'mountain dew' when he arrived back at PSV. Otnel asked, "What is this?" Jeff told him it was the Mountain Dew he asked for. This went back and forth a bit with Otnel not quite understanding and Jeff not quite realizing what the confusion was. Jeff explained the whole story of how he bought what Otnel had asked for.

Unfortunately, Jeff did not know that Mountain Dew was the name of a soda. What he had purchased was bootleg liquor . . . a whole case of it! Otnel has never let Jeff live that one down and even today, it took Otnel and Goatie about fifteen minutes to stop laughing; just the mention of the words 'mountain dew' sent them into hysterics!

More fun and games happened every Wednesday night which was 'Jump Up' night. Everyone would meet at the bar and because it was so popular and crowded with people, there were sometimes guests and yachties three rows deep. Crews from boats would bring coolers with their own liquor in them, keeping them on the other side of the wall by the pavilion. During these crowded evenings, guests would have to eat in shifts, three different ones actually, to accommodate everyone. Boats came from Palm Island, Bequia, and Mayreau to join in the fun and there was music and games. Goatie and Otnel remember the Queen contest held which makes them just laugh when asked anything about!

In addition to these particularly fun times, Mr. Richardson made working every day a pleasure on Petit St. Vincent. He would sit around and talk to staff often, just to be friendly. If he needed to speak with someone individually, he would take them aside and have a private conversation by the cannon. Jennifer thinks there must have been "hundreds if not thousands of conversations at that cannon."[7] He always asked the staff what they thought and really listened to their answers. Haze would even write a letter to each staff member at the end of the season and put with it a bonus for the outstanding work they had given to PSV over the year.

Otnel was very close to Mr. Richardson, just as every other person working on the island was. And he said how complementary Haze was all the time to the staff members. One day, after Hurricane Ivan of 2004 hit, Mr. Richardson told Otnel, 'Diego, you are a tower of strength.' That comment, along with the letters and compliments Haze handed out, meant a lot to him. Otnel still keeps the letters that Haze sent. Mr. Richardson was also kind enough to provide staff trips from donations which were given by the guests as well as monies he contributed himself. Otnel was able to accompany the other staff members who had been chosen to go as a reward for excellent service. He went on almost all the trips and was able to travel to such wonderful destinations as Puerta La Cruz, Las Vegas, and Miami, just to name a few.

Mr. Richardson would go away during the off-season, which originally was May through August. One time, when the end of season was drawing near and it was time for Haze to return, Otnel received a phone call from him saying he would not be able to be there on opening day but that he was confident Otnel would be able to take charge and handle everything himself. It was later revealed that Mr. Richardson had surgery while he was away and that one of his lungs was removed. Because of this, a swimming pool was added to their home shortly thereafter so Haze could use it as a therapy pool to help in his recovery.

Many employees were provided uniforms to wear every day. The uniforms the wait staff wore during the regular open season were made locally. Otnel remembers going to Petite Martinique and working with Fedelin Bethel [remember she is Lily's Ollivierre's daughter-in-law] who had been hired to make these garments as well as other beautiful items such as the napkins and the drapery in the cottages. The photograph of Otnel shows the sailboat design that was worn for quite a few years. All the material used was purchased in Bequia then given to Fedelin for sewing. She made many lovely garments for all the staff!

No matter how many projects had been worked on in the past, most every year there were additional large ones to complete such as: dock repairs, upgrades to the restaurant pavilion, a roof put on the newly constructed wine cellar, replacement of any rotting wood on the buildings, and varnishing underneath the new wood, just to name a few of them. Haze would make plans available by March so Otnel could get in needed supplies, especially if they wanted greenheart wood from Guyana, South America. If the order wasn't placed quickly enough, the Mustique Company would soon place their order and take most of the available lumber for their own building needs.

Otnel sporting the old 'sailboat' uniforms standing by the famous cannon

Other improvements and changes over the years included installing a new staff kitchen, expanding the laundry room, and adding three new staff quarter's buildings. Originally there were four housing units for the staff members before Otnel came onboard. Additional duties Otnel had included 'getting back to the basics;' Kate Nichols would sometimes request that Otnel and walk around the island with her so she could show him which bushes to trim. As she pointed, he cut. Otnel said he was always requested because "it wasn't Goatie's thing!"[8] He said that although the Nichols weren't there very often, he enjoyed the time they were able to spend on the island with the staff!

Otnel Samuel has been an integral part of the team at PSV and was instrumental in running the daily activities along with Mr. and Mrs. Richardson. He is an invaluable asset to the island in so many ways they are indescribable. He alone is responsible for many vital roles and continues still today as assistant manager, contributing his talent, abilities, and attention to make Petit St. Vincent such a sought after destination.

As Theodore Roosevelt pointed out, "Never throughout history has a man who lived a life of ease left a name worth remembering."[9] Noel Victory and Otnel Samuel have gone above and beyond in their life's dedication to PSV. Both DEFINITELY have a name worth remembering.

CHAPTER EIGHT

THE WOMEN OF PSV

In addition to all the men who were highlighted in the previous chapters, the important contributions of the women of PSV should not be overlooked. There were in the past forty plus years many women who worked diligently each day towards the success of the island, and without them the resort would not exist. The island is grateful to them and to the hardworking women who currently work in the office, spa, boutique, and the restaurants, as well as the kitchen, laundry, and housekeeping staff. They have demonstrated incredible dedication to their positions for which the guests and management are extremely appreciative.

Although every individual's contribution is essential in making the island what it is today, there are a few key women whose roles were instrumental in the creation and history of Petit St. Vincent as well as its continued success. They should therefore be acknowledged. Four individuals stand out in this regard.

JENNIFER RICHARDSON

Jennifer Richardson met Haze in the early 1970s when she was working as a yacht charter broker in the United States. She was enamored when she saw this handsome six-foot blond! Not too long after, she gave up her life in Florida and joined Haze on Petit St. Vincent. In 1976, they flew to the states and married in beautiful Sugarbush, Vermont. She called him 'Prince' and he called her 'Little Flower.'

A funny story Jennifer recalls is when Haze brought her to the island for the first time. He flew to Martinique to pick her up in his NA Piper Apache airplane.

They always called it "N Triple 2, 4, Pappa," which was the tail number for the plane [N2224P]. It was about an hour's flight and just as they were descending to make their landing in the Grenadines, all she could see was water, waves, a reef, and a flat brown and green area. Where was the airport, the landing strip, she thought? She was very scared, but trying to keep her cool, she very quietly asked him where they were going to land. Haze calmly replied, "You see that cow right down there in the middle of that flat area? As soon as he moves, we're going to land right there."[1]

Sign on Union Island's airfield

When she finally saw the island for the first time, she was astounded. How could this be created by a man who didn't have experience in many of the fields necessary to create the island we all know and love? Plumbing, electrical, construction, and engineering knowledge were all necessary . . . and the list never seemed to end. Not only was he obviously very talented, she also discovered very quickly that Haze was very well respected, not only on PSV, but also on the neighboring islands. "In all the years they were together, Jennifer can truthfully say she never knew of any one person who didn't like him."[2]

Haze on his way to Martinique for cheese and wine

Haze's plane, the N Triple 2, 4, Pappa

They each brought different strengths to the island so the responsibilities that weren't necessarily his forte, Jennifer would handle and vice versa. She believes they were the perfect team. Since they were both from the New England area of the United States, Jennifer feels they just 'knew how to figure things out.' And they were also both perfectionists! Although Jennifer loved the island from the very beginning, she knew there would be ways of living that would be different from her hometown. She used her creativity to accent the resort by designing different decorations such as putting flowers in conch shells for the dining room tables. Ms. Stewart, a local lady on Petite Martinique, made placemats for the luncheon meals and took fabrics Jennifer gave her and created a variety of table linens so that each evening the dining room could have a different look. Later, Fedelin Bethel, Lily Ollivierre's daughter-in-law [as mentioned in Chapter Seven] was the official seamstress of the island who made many items including curtains and uniforms. Jennifer had to learn to work with nature and with what was available. "You just had to think outside the box," she admitted.[3]

She was constantly asked how she could live on such a remote island for so long, and she always answered, "I guess if Haze was working on the Alaskan pipeline, I would learn to love living in Alaska. Wherever he was is where I wanted to be."[4] If she had to make a few slight adjustments to how she went about her day compared to living in the United States, so be it. And changes, she found out, were slightly more than she anticipated. Early on, Jennifer remembers using an IBM Selectric typewriter and one day it just stopped working. She told Haze it needed to be repaired as soon as possible so she could continue her work. She was amazed when he pointed out that a man would have to come from Barbados to fix it . . . and that would take about two weeks!

A friend of the Richardson's was coming down to visit one year and they were very excited for her arrival. This friend knew the island had limited access to many things, so she asked Jennifer if there was anything she could bring her from the United States. She thought the Richardson's might want special chocolates or other delicacies they couldn't get in the islands. Jennifer's reply was, "Please get the largest suitcase you can find and fill it with light bulbs!"[5] The friend thought that was a crazy request, but the customs men thought it was even crazier when they opened up the suitcase and found one hundred light bulbs!

As you can see, they sometimes had to go to great lengths in order to get the supplies they needed. There just was not the availability of goods on the neighboring islands. They often ended up flying to the States with Haze having a suitcase full of broken parts and pieces of unrecognizable mechanical equipment in tow. So of course, a similar situation occurred then they entered customs. The officials just had to ask what all the 'junk' was and why they felt it necessary to bring it into the United States! Haze had a lot of explaining to do and sometimes he would have to pick up each piece and say what it came from and what was going to be fixed on it to even get through. When the officials were finally satisfied, there was no time for rest; they both would go off in different directions and shop, shop, shop. By the time they were ready to return to PSV, they were exhausted and needed a vacation!

Dealing with Vincentian customs always ended up as an interesting situation they had to contend with as well. But Haze usually had a plan. Jennifer remembers a few times, when they just didn't feel like taking forever to get through the red tape, Haze would time it so that he was circling the airport at about 5:24 pm This was because if sunset was at 5:30 pm, for example, they had to land no later than that or the airport would close. But Haze also knew that the officials were anxious to leave after a hard day's work and hopefully that would get them through a bit quicker! Jennifer was always prepared with a 'donation' of National Enquirer magazines which many of the custom's workers loved she observed. "They always asked me to bring some back for them, which I did of course . . . after reading them first![6]

One year, Haze and Jennifer bought a new labrador puppy and wanted to bring him to the island to join their already large family of five dogs! But St. Vincent had a quarantine law that would make him remain in the government's care for six months. They knew this would not be possible for them to comply with. So Haze wanted to know what they should do about it; he asked Jennifer how they will get the dog through customs. Jennifer told him not to worry, that by the time he arrived with the puppy, she would have a plan figured out. So Haze loaded up 'Alexander' and flew down to Union Island. Jennifer, Chester Belmar [the boat captain], and Sampson, Delilah, Brandy, Clarence, and Cleopatra [the resident Labs] got on their boat, *Wakiva* , and headed there as well. When they arrived at Union Island, they got off the boat to meet Haze as his plane arrived. When he exited the plane, Jennifer told the dogs, "Dad's home, dad's home, go get him!" As the dogs excitedly ran over to him, he opened the door to the plane to grab his luggage, and Jennifer called the dogs back so they could all go home. Sampson, Delilah, Brandy, Clarence, Cleopatra, AND Alexander all returned to her! Six yellow labradors . . . happy to be heading home!

Chester stands in high regards with the Richardson's. Jennifer described him as a very handsome man who "could take a boat through the narrow channel by Mopion Island with his eyes closed."[7] Even though Jennifer has sailed many times, she was never allowed to sail through that tricky channel.

Private picnic on Mopion Island

The little island of Mopion, which can only be measured in feet rather than acres, lies just north and west of PSV, and has a bit of history itself. It is also named Petit St. Richardson because PSV maintains it and sports a nice palapa where guests can relax. Jennifer remembers friends of hers, including a New York photographer and his Ford Model girlfriend were married there, which she was a part of, and there were many fashion shoots done there as well by magazines such as *Vogue*. One couple accidentally stayed on the island overnight and survived! Guests can be dropped off on the island for a picnic lunch to enjoy their time together. They are there for a few hours of the day and then a boat comes back to pick them up later. A little slice of heaven near the bigger piece of the pie!

With all of the work, time, and effort that Jennifer and Haze spent running the island, it was the staff that kept them going. They both cared for each and every one of them very much. "Bob and Marge, John and Casey, and Slick the chef, were one of the best staffs we ever had, many of whom had been there for years," declared Jennifer.[8] Of course, the goal of the island was to make sure the guests' needs were met, but their hearts belonged to the staff and they continued through each day because of their love for the people they worked with. Haze ingrained on the workers the importance of every detail of how each guest should be treated, but Jennifer saw that this caring feeling came instinctively to them. It was an attitude they carried naturally each day . . . the training was just in the job details they had to perform.

In order for the Richardson's to gain a guest's perspective, they would sometimes stay in the cottages from time to time. This allowed them to decide what features needed to be added, which ones worked well and what, if any, improvements needed to be made. Haze was a stickler for details but that is probably because he was so smart and ahead of his time. And this character trait showed in many other areas as well. Jennifer was amazed that Haze was interested in using reverse osmosis for water purification before most people even knew what that was. She laughs when she remembers how Haze would walk around the island with his 8 ½" x 11" yellow legal pads where he kept all his information and notes. No matter where he was, at his desk, at the bar, or anywhere, the pads were there with him. He was known for that!

Being a former yachtie himself, Haze always encouraged the yacht's visits and was always very gracious to these visitors. He wanted them to look at the accommodations and amenities, tell their friends, and maybe even book a future stay! The only time island visitation was discouraged was when twenty to forty people would arrive on a head boat, which was a large sailing boat which took tourists to the islands on day sails, usually out of Union Island. They would take over a large portion of the island and the beaches which made it uncomfortable for the paying guests. Other than that, Haze welcomed regular yachts, encouraged them to anchor, and possibly book for dinner.

When the country of St. Vincent and the Grenadines had finally gained its independence in 1979, Haze and Jennifer were invited to meet the Queen of England at one of two events: a cocktail reception on board Her Majesty's Yacht *Brittania* or a tea with Queen Elizabeth at the then Prime Minister Sir James Mitchell's hotel, The Frangipani, on the island of Bequia. Although Haze could not make it, Jennifer decided to attend the tea and joined Sir James Mitchell and others for this exciting occasion. She contacted a representative at the British Embassy in Washington prior to the tea to obtain proper etiquette rules to follow when she met Queen Elizabeth.

"I remember holding the phone in one hand and practicing my curtsy at the same time, listening to the rules which were as follows:

Rule #1 - *As I took a deep bow, the representative said, ". . . and since you American's tend to overdo everything, just a slight bend at the knees will do."* Because I was in Maine at the time and it was quite chilly that day, the appearance of me dressed in an old t-shirt sporting sweat pants was probably not quite the look they would be wanting either I thought!

Rule #2 - *You must wait for the Queen to speak. If she does not speak, you may not speak. If she does speak, then follow with Rule #3.*

Rule #3 - *Answer the Queen's first comment with the words 'Your Majesty' at the end of your sentence.*

And the last rule:

Rule #4 - *If the Queen speaks to you again, you then follow your sentences by addressing her as Ma'am."* [9]

Jennifer was very excited and when the day arrived, she was in Bequia dressed in a beautiful suit, sporting black patent leather high-heeled shoes, and a hat more suited for Joan Collins, adorning her head. "She more than likely expected me to be in a sundress or something more appropriate to where I was living at the time," thought Jennifer later.[10] It was a warm day which unfortunately produced a 'sweaty Jennifer' and her shoes presented a 'hobbling Jennifer' each time the heels sunk down into the grass which was unfortunately with every step she took toward the Queen. It was finally her turn to meet Queen Elizabeth and she was ready. With Rules #1 through #4 having been properly memorized, she walked up to Queen Elizabeth, stuck out her hand and exclaimed "Welcome to the Grenadines!"

"Well, I blew it I thought," declared Jennifer.[11] And as she was sulking for the mistakes she made, a little ray of cheerfulness brightened her otherwise unhappy feeling. "Prince Phillip came up to me after the receiving line and chatted for a very long time with me, no doubt out of sympathy. He was charming and witty," remembers Jennifer warmly.[12] She declared the event was something to be compared to "meeting the postage stamp"[13] because Queen Elizabeth just looked at her and didn't utter a single word!

Sir James Mitchell, Colin Tenant, HRM Queen Elizabeth, and Jennifer [in the background]

Jennifer believes the best days at PSV, the 'glory days' if you will, were during the time Haze was a part of the island. "All the stars lined up to create a great environment."[14] Petit St. Vincent reached a pinnacle level which she thinks will always remain as credit to this wonderful man. "It was a very special period in time," Jennifer fondly remembers. "The staff and the management were almost family, all working together to create a wonderful experience for the guests. The guests loved the staff and the dogs loved the guests. Many of the guests even became personal friends whom I am still close to this day . . . it all just worked . . . it truly was a magical time."[15]

Running an island, which is really just a rock in the middle of the ocean, took its toll on this wonderful team. "There were no holidays or times off the island in those days for us . . . we focused all of our attention on the guests and the staff, and unfortunately this left little time for us. While I know we made countless happy memories for countless people, we didn't know then to take time for ourselves; a failing we both regretted," Jennifer confided.[16] Although they divorced years later, she never remarried and has always held Haze close to her heart and always will.

LYNN RICHARDSON

Lynn Richardson was married to Haze in the 1990s and although there are many stories from the time she was a part of the island and from the years of commitment she put into PSV, she wishes her story remain private. Therefore, no additional information will be written about Lynn or her time with Mr. Richardson as she requested.

MATTIE BETHEL

Matilda Bethel, 'Mattie' as everyone calls her, came to PSV in 1973. She began work in Carriacou at the age of 16 ½. She was a telephone operator for three years until one day Haze Richardson walked into her place of employment and mentioned he thought she looked familiar. Mattie told him that although she never had the opportunity of meeting him, her two brothers, Chester and Reynolds, worked for him. He asked if she would like to join his staff and the rest is history!

She left her telephone job and began working in the boutique on PSV. Mattie enjoyed her work there under Jennifer's guidance. "She was a very loyal, hardworking person who took the place to a high standard," Mattie acknowledged of Jennifer.[17] She organized some wonderful changes to the boutique when she arrived and contributed to PSV in astonishing ways. Mattie believes Jennifer "helped to make the island the best in the world!"[18]

From left to right, Mattie Bethel, Cora Herbert (behind counter), and Lee Blissett

Mattie worked on the island for seven years until she became pregnant with her children. She stayed home to raise and nurture them but when the first boy was eight years old and her second child reached school age of five, Haze immediately called her back to work! This was 1989 and there was an opening available. She returned to her daily schedule in the boutique from 9:00 am to 5:00 pm, but in just a few years, she was promoted to the position of bookkeeper and accounts clerk where her duties then entailed reconciling the daily cash, making out vendor checks, and doing the payroll and banking.

For this position, she worked from 11:00 am to 7:00 pm. But the last two hours had nothing to do with office work. Mattie became the island's official dog walker! She confessed this was the best part of her day. Haze was a very big animal lover and had a total of seven dogs at that time, all labradors. They were not only a wonderful part of PSV but a very big part of Mattie's life as well. During the workday they could be found lying near her desk on their dog beds and when it was 5:00 pm, they knew it! That was play time for them, and Mattie and the dogs took off on their adventure. They would walk on the beaches, play in the sand, and greet the guests. When dinner time rolled around, Mattie would take the dogs to the Richardson's house and feed them. They knew when it was time for that as well!

Famous Dog Walker, Mattie Bethel

Chibby, Minnie, Hercules, Delphi, Mr. Green, Zeus, and Hera were some of the dogs that would greet the arriving guests. The labradors were a very important part of the history of the island and many returning guests ask about them. They felt the dogs added a homey feeling to PSV and missed them when they were not there. Sometimes guests would 'adopt a dog' for the day and they would return him or her in time for supper! But it was not all play for these dogs; they also had duties to fulfill . . . they guarded the island from unwanted creatures and Chibby is credited with taking on an octopus and winning! They kept the island tidy by picking up fallen coconuts and chewing on them but unfortunately one day Zeus ran into a coconut tree when he wasn't watching where he was going. Mattie says, "He was never quite the same after that sadly."[19] Such is the life of a dog.

Problems would arise sometimes as any dog owner might expect. Chibby decided he was going to be the island bully. He took it upon himself to dislike most staff members and he shared that feeling one day by biting one of them. He also decided to do a very stupid thing . . . he bit Haze as well. So as any good dog owner would do, Haze admitted homeschooling defeat and hired a trainer who flew in from Florida to work with the mischievous dog. He explained that Chibby must learn his manners and the way to get him to stop hating the staff was to have one of them, other than Mattie, walk the dog. The trainer believed the color of the person's skin seemed to be the trouble. Because Saga was the darkest-skinned person on the island, he was unfortunately the chosen one. His days were normally spent in the workshop but he was now promoted to the coveted position of taking grumpy Chibby for a daily walk. That must have been a hard sell!

Maurice Roche, boat captain, displaying Hera *flag*

So he walked the dog every day and with time, understanding, and a very patient Saga, Chibby became the nice, friendly labrador everyone else knew and loved and could finally take his daily stroll with his head held high sporting an air of self respect! But Chibby could only withhold his mischief for so long. It was rumored that he swam to Mopion one day and before he could be rescued, decided to swim back! He had Mattie worried sick. What a show off!

Two boats on the island have received their names in honor of the dogs and both are still being used today . . . *Hera* and *Zeus*! Mr. Green had the dining room named after him instead of it being called a boring but obvious name of 'top dining room.' Unfortunately, poor Chibby was not immortalized in any way.

Haze relaxing on the dock with a few doggie friends!

Dogs were not Mattie's only animal love; she also voluntarily took care of a couple of Brazilian parrots who arrived one year on a South African yacht. Their cages were kept by the bar area. Apparently one parrot, who shall remain nameless, did not like Elvis King who was one of the waiters. To make matters worse, a bartender named Eric trained the parrot to say terrible things about him; he would squawk his head off every time Elvis would walk by! Although they were quite happy in their cage, someone decided to open the door one year so they could have their freedom. Although Cedric and Frederic stayed around for quite awhile and seemed happy on the island, they suddenly disappeared one day. Allegedly, they flew away to the neighboring islands but some believe their fate was determined by the dogs. No matter which story holds true, they had been a wonderful contribution to PSV.

Mattie caring for the parrots

Parakeets also made Petit St. Vincent their home and, of course, Mattie was the one who took care of them! They lived in a cage kept in the office and their names were Romeo and Juliet. If that weren't enough animals, a sea turtle project was begun in 2008. A pen was constructed off the beach by the dock house to enclose hawksbill and green sea turtles. It was originally planned as a way for guests to be part of watching their eggs hatch and participating in the hatchlings' release. It was met with mixed reviews and difficulties, so when the new owners' bought the island, the pen was taken apart and the turtles released.

Land tortoises currently live on PSV as a gift to the new owner's son, Jack. He brought the first one with him from Canouan on his yacht *Galileo*. They have a small pit-type enclosure by the reception area. It used to be the area the dogs loved to lay in to bask in the sun. Mattie takes the restaurant's leftover fruits and vegetables from the night before and painstakingly cuts each piece into the small size needed for the turtles to chew. She feeds them every morning when she arrives to work and even stops at a particular tree to bring up fresh leaves which functions as their food-serving platters. These are some well taken care of turtles! And of course they are a big hit for the youngsters visiting the island. The first family of turtles grew too large for the pen so they were released on the island. They have been spotted by various guests crawling through the brush and trees and have been reported as thriving on their freedom.

Mattie speaks affectionately about her time with the dogs and other animals but that is not the reason she has stayed with PSV for so long. It was Mr. Richardson that made her want to continue. She felt he was such a good employer to her and the other staff members. "He was born to be a boss and owner," she says.[20] However, he was very strict in some ways by insisting everyone tell the truth. No matter what a person might have done, or what mistakes they may have made, Mr. Richardson wanted honesty from them. That way the issue could be resolved and everyone could move on. He did not hold grudges or bring things up later. "He was very considerate," Mattie acknowledged.[21]

"I learned a lot from Mr. Richardson," Mattie admitted.[22] She remembers one day when she was having a particularly frustrating time with the books and was making mistakes, he told her, 'That is why there is a rubber thing on the end of your pencil. Don't worry about it. Just do it better the next time.' He was very patient with her and never uttered a bad word, even if she knew he had more pressing things on his mind. And always, when he had to correct someone, he would start with a story of some kind! Of course it eventually led to the issue at hand but was a great way to begin an uncomfortable conversation. When she was worried about taking the bookkeeping job because she had no experience, he was quick to encourage her and she can still hear him saying, 'Mattie, you can do it . . . just take one day at a time.' Like many others on the island, Haze would hire an individual to do one job but quickly promoted them, giving them the encouragement they needed to succeed in another.

Mattie was very thankful for his generosity as well when it came time to send one of her children to school. She asked for a loan and it was granted. Each month a small amount would be taken from her check. She was very grateful because without that money, she would not have been able to afford his education.

Mattie is a multi-talented person who has honored the island with her continued presence all these years. She has dedicated much of her life to PSV and for that everyone is thankful. She has created a working environment which is fun and exciting and the island is appreciative of the way she has handled the responsibilities that have been entrusted to her. Every year new challenges were set and every year she accomplished them. Mattie continues to be one of the highlights when guests arrive . . . a smiling face and a warm heart are the qualities she possesses and those are what she shares with everyone she meets.

DELIA CUFFY

Miss Delia Cuffy is another important and instrumental woman on PSV. She came to the island twenty-two years ago from St. Vincent. A friend referred her to Mr. Richardson and when she arrived, she was employed immediately and went to work in the staff kitchen. She stayed in that position for about one year.

It was Delia's job to make sure the area was clean and tidy and to serve the staff their meals. Each evening she also baked fresh bread for them so that it would be ready when they arrived for breakfast. Early in her employment, she worked with a chef who she claimed only wanted to be responsible for cutting meat and felt women were not able to do that type of work but should stick to cleaning and serving. So one day, she got fed up with that arrangement and took over! She cut the meat and let him serve and clean for the day! She also wanted her views to be heard so she pointed out to him she thought he made too much food altogether and that it was very wasteful indeed!

They usually served the staff three meals a day which consisted of bread, rice, and one kind of meat or another. They would get chicken legs, wings, goat, and fresh pork from St. Vincent, and turkey from the United States. They would also make a very nice fish broth for those who preferred a vegetarian meal! She remembers Haze telling her what a good job she had been doing so he decided to move her to the main dining room when a position came available. Delia feels it was very nice that Haze was always quick to promote.

Her new title was assistant pastry chef and she worked under a man named Marcus who was the head pastry chef. Many return guests will fondly remember him. She liked this job much better because she was learning something new every day. He taught her how to bake cookies, muffins, tea bread, and cakes all from scratch for the guests' breakfasts. Because Haze was always looking to improve the island, he later hired a Belgian pastry chef who taught her bake with chocolate and how to use it in many recipes. Delia thought he was a very good teacher and the guests were very pleased with the results! When he decided to leave however, Delia was pleasantly surprised that Marcus was moved to the position of head chef and she was now promoted to head pastry chef!

Haze began a routine of bringing in guest chefs to improve the staff's training in new cuisines. Every time a chef would come in, Delia and the staff were taught new ideas, tips, and recipes to follow and when the chef left, he took with them his favorite recipes from the island! Delia remembers being fortunate enough in 2006 to accompany Trevor, another chef, to the United States where he worked in the off-season. She learned new ways to work with bread and she in turn taught them how to make her famous Guinness ice cream! "Haze was always improving on everything," Delia observed.[23]

She recalls how much the guests loved the meals, especially after a barbecue pit was built in the Mr. Green dining area located above the pavilion. They would cook all the meats on the grill including an island favorite, blackened fish. They would also offer a variety of different cold salads, rice, breads, and desserts. Guests would gather for lunch and have a great time talking and enjoying each other's company. She noticed times are somewhat different today however; many guests stay in their cottages more often than they used to and order foods to be delivered to them.

In 2009, Delia started having problems with her hands; they were swollen and the joints were inflamed. The doctor explained she was experiencing these problems because of the continued years of work she did with her hands. She was therefore offered a position in the storeroom where she now has the daily duties of keeping track of the island's food inventory. Most of the foodstuffs still come from St. Vincent and only if items are not available, they will be ordered from the United States.

Besides her regular job, Delia was responsible for taking care of the dogs when Mattie was off. She took pleasure in helping out in this way because everyone lived as one big family and looked out for one another. But her happiest times were when 'Sports Day' would roll around. Delia was the organizer of this event and it was a day where the staff could relax, have some fun, play games, and just loosen up. Many of the guests over the years would join in too! It was great fun and everyone looked forward to it!

The first Sports Day was held in 1992. Everyone was presented with a commemorative t-shirt as a delightful symbol of the day. These were also given to guests who chose to participate in the events as well. There were many contests like the egg and spoon race, the needle and thread, and the three-legged race. There were kayak races, Sunfish sailing contests, and water balloon races. There was also a relay race around the entire island. A favorite and frustration for many was the game where you had to take a bucket of seawater and move it to an empty bucket with a 'holey' spoon! In the evening, the fun continued with a domino competition! Sport's Day was quite important to the staff and was about the only day they could all have fun sharing together in each other's company.

Each and every woman of PSV has given so much of their time and talent to insure the success of the island. Without the special gifts each one has so graciously shared, PSV would not be what it is today. The old phrase 'a woman's place is in the home' should be long forgotten as it certainly does not apply to these women who have proven that not only can they manage a career, but they do it with competence, style, and a little 'extra pizzazz.' Petit St. Vincent honors these women for their long years of dedication and hard work.

CHAPTER NINE

SIGNIFICANT OTHERS

JAMES MITCHELL

When the island was at its beginning phase, Mr. James Fitz-Allen Mitchell was starting to represent the people of the Grenadines as the Minister of Trade, Agriculture, Labor, and Tourism, a position which he held from 1967 until 1972. Because of his continual career in government in the Grenadines, spanning nearly four decades of service, he remembers many stories associated with Petit St. Vincent during his time as Premier and then Prime Minister. He was bestowed the honor of being knighted by Queen Elizabeth II in 1995.

During the time Mr. Nichols was developing the island of Petit St. Vincent, others were also beginning or in the midst of constructing their own paradises such as Colin Tenant on Mustique and John Caldwell on Prune Island, now called Palm Island. There was no electricity, phones, or roads of any substance nor were there any airports at that time. Sir James is from the island of Bequia where he still currently resides. He decided to build a hotel around the same time which opened in 1967, called the Frangipani Hotel. It started with five rooms which had lanterns for lights. Arne Hasselqvist designed some of the rooms at this hotel as he did on Palm Island, Petit St. Vincent, and finally Mustique.

Because the local communities were mostly families of sea farers, many of these new establishments were searching for workers in St. Vincent because the people of the area were known for their excellent masonry and stonework abilities. Their skills had been refined over the years and these men produced remarkable, as well as quality, construction. The same situation occurred when it came to finding skilled chefs. There were just none around the area so businesses had to bring in qualified candidates from other parts of the globe. But through the years, many of the local residents were able to learn from the expertise of others and were thus able to take over the positions themselves, which allowed for a more balanced choice of competent applicants.

The different islands all had one thing in common however. They all prospered on tourism brought in by guests and especially the yachting community. Because of this, Sir James, Haze, and a number of other key people on surrounding islands came up with a plan to make the best use of the people sailing through the Grenadines. They each chose a night to have a barbecue, so as the yachts continued their journeys in the West Indies, they could stop at each island and have a wonderful dinner and a few drinks. Mustique chose Wednesday nights at Basil's Bar, Sir James chose Thursday held at the Frangipani in Bequia, the Mariners Hotel in St. Vincent had their barbecue on Friday night, with Petit St. Vincent rounding out the weekend by having theirs on Saturday. It was a good time for all and this arrangement worked out quite well for many years.

Sir James became Premier of St. Vincent and the Grenadines in 1972 and soon after arranged a meeting at PSV with the other Premiers of the area including Eric Gairy of Grenada and John Compton of St. Lucia. The discussion regarded the issue of workers from the Grenadian islands traveling to and from the St. Vincentian islands. This had been an ongoing concern because many of the workers employed on Petit St. Vincent lived in Petite Martinique and therefore were supposed to obtain work permits to do so. Additionally, workers usually wanted to go back home to their country when they had time off, or in the case of PM, they would go home every night.

After spending the weekend on PSV, the three men devised a plan and called it the Petit St. Vincent Initiative. It was the first time in history freedom of movement among citizens of Grenada, St. Vincent and the Grenadines, and St.

Lucia was allowed where no work permits were necessary to be employed on one island or another. The plan was agreed upon and it was executed by Sir James.

Another project which Sir James worked on with Haze was how to handle the issue of the fisherman living on PSV's beaches in a fair and equitable manner. A number of PSV employees were involved with one of the initial meetings, finally arriving at a solution to build a more permanent structure for the fishermen to live. Sir James recalls how he later went to PSV after that discussion to help decide how the accommodations were to be allocated for the fishermen and under what terms. However, during the later part of June and the early part of July 1987, Haze requested that someone from the government help resolve another issue as well. Financiers were to be visiting the island around that time and Mr. Richardson requested that the fishermen be relocated for the dates the visitors were to be on the island.

There were two different problems that needed to be addressed and both issues were on the table simultaneously: How and where to relocate the fishermen for the time the investors were visiting, and finding a long-term solution for a more permanent residence for these men. Issue number one was resolved in a few hours; the fishermen were to remove their tents from the island and were provided a rental home to stay in on the island of Petite Martinique for a few days. The agreement followed that the management of Petit St. Vincent would incur all costs of this temporary relocation and that they were allowed to return with their tents at the end of this time period.

The second issue was a little more difficult to resolve. According to Chief Fisheries Officer Kerwyn Morris' report dated July 27, 1987, who officiated at this meeting:

"It was not an easy task mediating in the two joint sessions that were necessary before a final position was arrived at. It must be remembered that both sides have not been

on the best of terms for several years and several unpleasant events were recalled during the meetings." [1]

The fishermen held the position that the structure Mr. Richardson was going to build for them needed to be adequate for no less than fifty men and that wherever the building was located, it must allow for an unobstructed view of the harbor. Management's position dictated that no more than nine men should be accommodated, they were not going to remove any vegetation from the island to permit an unobstructed view, and that they believed an alternate location on another island such as Canouan or Petit Tabac would be in the best interest of Petit St. Vincent. It was noted by the presiding officer that the Honorable Minister of Trade, Industry, and Agriculture had set up a deal with the Mustique Company on a similar issue and their resolution was to provide accommodations for seventy-five fishermen.

The meeting was at a standstill and therefore was reset to resume the following Sunday. Unfortunately, time did not have a way of reconciling the issues; when discussions resumed, the talks heated up with both sides calling fouls against each other, with not much more than minor issues being resolved during the day. Haze's point was that of wanting to support the government as long as it did so in a way as to not detract from the operation of the resort. Although no permanent solutions were arrived at that day, eventually a residence was built and the fishermen and the resort lived harmoniously. Over subsequent years, the fishermen moved on to other areas; when their market of selling fish to Martinique closed, when they simply overfished the areas, and when some fisherman passed away.

Jennifer, Haze, and Sir James at opening of the Fisherman's Camp

Taking time off from his fifteen hour work schedule Sir James was always present at PSV and the Bequia Easter regattas which he subsequently created. He brought along distinguished friends like Sir John Compton of St Lucia in their yacht but never won a race, he recounts.[2] The PSV regatta attracted many Trinidad boats during the oil boom days of the seventies but came to an end with attraction to American visitors returning for Thanksgiving.

Another particular memory Sir James enjoys associated with PSV was the year he stayed on the island so he could watch history in the making. It was June 1994 and he had successfully secured the installation of electricity in Canouan. From the northern side of Petit St. Vincent, he was able to watch the island of Canouan finally receive electrical power . . . the lights were switched on!

On another particular visit to the island, the topic of septic systems, of all things, came up! Quite an interesting conversation to be having with the Prime Minister, Haze must have thought. However, Haze wanted to ask Sir James' opinion on how to keep the problems he was having of clogged septic systems at bay. Because Sir James has a degree in soil bacteriology, he had a solution. He shared the fact that the difficulties arising were from not only a paper problem but told Haze he must also stop using chlorine because it poisons the bacteria which keeps things working properly. But the second most important instruction he must follow . . . put horse dung into the septic system!

The directions were simple enough. Each spring and autumn, a handful of dung should be put into each toilet and flushed. After the trouble seemed corrected, this only had to be done once per year. There was one drawback with these directions however . . . where were the horses to supply the needed magic potion? There were certainly none on Petit St. Vincent. So who should they call to resolve this problem? Who else but Arnie Hasselqvist, the architect!!! He got right to work and shipped the dung to Haze from Mustique. Problem solved!

In the early days as mentioned, Sir James remembers how Haze was always very generous to him; he would often invite him to visit the island where he stayed in one of the cottages. They would dine together and in return, when issues arose on PSV, Sir James would take his calls. However, all that changed when a problem surfaced regarding PSV during his tenure as Minister of Finance. Sir James explained that each island has a different tax arrangement with the government. Some have flat taxes, occupancy taxes, or tax holidays for certain projects, and/or income taxes as well. Some islands such as Mustique and Canouan are especially known for their fine support of charitable trusts along with all of those. If a tax holiday is granted, it needs to be renegotiated every fifteen years or at that time, the regular tax agreement goes back into effect.

Around the year 1998, a discrepancy surfaced with the figures Petit St. Vincent was claiming which were turned into the government. Because of published statements obtained from immigration records, the government knew exactly how busy the island was. A dispute with the tax department therefore ensued when the gross income of PSV was not equivalent to the occupancy which was officially recorded. PSV manager had difficulty responding to tax authorities' incompatibility between gross revenue from arrivals' length of stay and revenue reported. Sir James recalls that Mr. Richardson contacted him about this problem but assured Haze that he could only set up a meeting to discuss the problem but in no way could influence the outcome. Unfortunately, Sir James feels this was the end of their friendship as Haze was not able to receive any additional assistance from him. The funds were never found.

Many demanding as well as uncomplicated times involving Sir James and Petit St. Vincent existed throughout the years but that is to be expected when there are so many people trying to reach a mutually-acceptable goal. In the end, collaboration and teamwork allowed the government and this island to work together for the good of both and much was accomplished over the years. PSV continues today to have a good working relationship with the St. Vincent authorities.

JOHN CORBETT

Mr. John Corbett was once a consultant who lived in San Francisco, California with his wife and three-year-old son. He decided to take a year or two off in 1977 and sail from California to the Grenadines. He never went back.

Racing Yacht Freya

John was the owner of sailing yacht *Freya,* buying her in 1970 with the two of them eventually logging over 120,000 miles together. [Phil Stephenson later bought *Freya* from him.] So when he arrived in the Grenadines, he decided to compete with her in the PSV Regatta. This is where he met Haze Richardson for the first time in 1978. He continued to enter each year, eventually helping Haze organize the race, and was even fortunate enough to win a few! He enjoyed the fact that they were always held during the resort's off-season as it was a time he, Haze, the yachties, and the staff could all relax and just enjoy themselves.

Haze and John became friends as a result of the regattas and after speaking to John many times and discovering the type of work he did, Haze decided to ask John for some assistance. Haze confided he sensed the staff was unhappy about something but could never quite put his finger on what was troubling them. He asked if John could help by checking with the staff himself. So that is exactly what

John did; he sat down and interviewed all eighty-plus staff members and was able to successfully get to the bottom of the confusion and unhappy feelings.

Haze had been very close to each and every staff member, even speaking with them, at least, on a weekly basis. John realized they cared for Haze just as much but it was this friendship and closeness which made them reluctant to share any complaints directly with him.

After John concluded the interviews, he explained his findings. Haze now understood why he had felt there was something wrong but could not pinpoint anything specific. Haze immediately wanted to know how to make things better. The issues included some housing problems and the fact there was no place the staff could gather to just relax. That same day, the two of them rode around the island on Haze's motorbike and picked out a location for a new recreational building. "Haze was always willing to act quickly when needed," John said.[3] Construction began later that week.

John also pointed out that it would be important for the island to have other heads of departments, since he felt it would be very difficult for Haze to continue managing all the staff as he had been doing. As John said to Haze, "you're just one person."[4] Haze agreed. However, when they went to set up a workable system, they found they had quite an arduous task at hand. Finding supervisors within the staff who already were working on the island was a problem. Most of the staff were from Bequia, Union, or Petite Martinique and they were all reluctant to take a management role. After some investigation, John found out that because friendship is very important to the West Indian people, they were afraid that being in a position of authority would put them in a bad light with their coworkers and friends. Therefore, initially they refused.

As John worked with them and explained how and why the selected staff would be able to be effective as a boss, but still could maintain their long-time friendships, they understood and agreed to take the positions. Although this had taken some convincing and nurturing, supervisors were eventually set up in various departments such as the garden, the office, housing and dining, and at the boat house.

Not only did the staff learn a lot during this time of adjustment but so did both John and Haze. They realized that because the West Indian people are naturally friendly, which was a trait Haze already saw in them, this attribute was also an important selling point for the island. They believed it was that particular quality the locals had which made many guests feel drawn to PSV. So by using this natural trait, they could achieve even more. They explained to the staff the amount of time they spent with each guest was very important; something that was expected. So for example, if a guest had questions about the garden, it was just as important to the island's productivity for the staff member to stop what they were doing to answer any questions or to give that guest a tour as it was for him to have worked on his task for that same amount of time.

John and Haze both liked this approach to managing people, John said, and realized that not subjecting any member of their staff to 'a plantation-like approach' was beneficial to all. "This was part of Haze's approach: It was the staff's job to make people happy by having lunch or whatever."[5] He always wanted there to be an easy, friendly relationship existing between staff and guests: Slick, the chef, would love to sit down and have a long chat with them, for example. John believes the reason so many people came back was this ease of communication. "They came back to see their friends; it was not just a vacation."[6]

All of this was made uncomplicated by Haze's natural personality; he was genuinely respectful and never 'put on airs.' "Haze was a kind and humourous man and a magnificent leader loved by his staff and guests alike, treating all with equal respect - whether a stranger, an acquaintance, a friend, or the head of a country,"

John remembers. "This feeling of mutual respect encouraged an easy, confident interrelationship between staff and guests, which became a hallmark of PSV."[7] John believes many became repeat guests because of just this factor. John speaks very highly of Haze but noticed one imperfection, if you could even call it that; he believes he was a bit of a procrastinator. But Haze was the first to admit this as he posted a sign on his office door which read, 'My decision is maybe . . . and that's final!' However, John admitted, "When action was urgent he acted courageously in a flash."[8]

"Beginning with the regattas a firm friendship grew between us, and over the next twenty-five years, I worked both as resort manager during Haze's absence and his business adviser on organization, staffing, and financial matters including Haze's acquisition of control of the resort," John said.[9] He sometimes stood in for day charters with *Freya* as well, and represented PSV on the board of directors of the Tobago Cays Marine Park.

Haze and John ran out of gas

John describes Haze as courageous and always calm, even in the worst situations. This characteristic turned out to be a lifesaver for both of them one frightful evening. John had just sailed back from Venezuela and found himself and *Freya* caught in a storm and aground on a reef. Haze had a PSV boat take him to John and they stayed aboard *Freya* all night long in an eventually unsuccessful struggle to help avert her from severe damage and sinking. John remembers they were in waist-deep water in the boat and were afraid the storm was going to completely demolish her. However, the night wasn't all spent in misery; since he had just returned from Venezuela with several cases of a popular Venezuelan beer, John jokingly divulged, "there were quite a few *Polars* floating around in the cabin conveniently at waist level which we just had to enjoy!" "Haze risked his life that night to help ensure my safety. That was extreme courage and generosity to a friend - not for the first time"[10]

During one of Haze's absences, which totaled about nine months when he had broken his hip skiing in Switzerland, John remembers how nice it was that Haze trusted him enough to give him full rein. "I would always have his full backing to carry out whatever he deemed necessary and authority and judgment was never questioned," recalls John.[11] He basically never called or checked in, knowing that if there were any issues, John was quite capable of handling anything that would arise.

There was a particularly funny occurrence that happened during the weekly cocktail party while Haze was off the island, John recalls. While talking to a guest, John suddenly found a hand caressing his bottom. He turned around to see the hand belonged to a lovely lady and when she saw him, her face turned beet red. "All she could say was, 'Oh my goodness Mr. Corbett! I thought you were my husband.' The only words I could manage were, 'if only!'"[12]

During his time on the island, John recalls other interesting incidents that arose . . . maybe not quite as personal as that one was though! Every week there was a barbecue that was held for the guests. On one of these evenings, a beautiful yacht pulled into the harbor and a crew member approached John, asking for permission for the owner, his guests, and crew to attend the party. John explained to the crewman that this event was strictly for guests and asked that PSV's apologies be conveyed to the yacht owner, who was apparently a very influential fashion and fragrance designer. The crewman was very insistent and, not wanting to take no for an answer, started peeling off US$100 bills one by one. John was polite but insistent that he could not make any exceptions. The man got very mad and left in quite a huff stating that PSV will regret that decision. As the saying goes, 'you can't please everyone.' Or would it be more befitting to remember the phrase 'money can't buy happiness!'

Another humorous story John recollects is when a couple came to spend their honeymoon on the island the day after their wedding. Just two days after they arrived, the groom walked into the office explaining explicitly, "Mr. Corbett the marriage is over, please have your staff book that woman on the first available Concorde back to London - and please extend my stay by one week." Although this put the staff in a funny place, the bride also shared her feelings that she wanted to leave and according to John, "she 'Concorded' out the next day! The husband, or shall we say ex-husband to be, stayed the extra week and appeared to have had a very happy time."[13]

One year, another out-of-the-ordinary incident occurred: there was a returning guest who decided to come to the resort as he had done the previous year. However, he wanted to advise the staff that a sensitive issue was at hand and that he would appreciate any assistance to avoid a potential disaster. According to John, the story from the man was relayed as follows:

"Unknown to his current wife, the first stay had been with his secretary whom he had introduced as his wife at the time. He had told his present wife that he had visited alone. As requested, I discussed this with the staff to try and ensure there were no snafus such as mentioning the 'wife' from the first trip. The secret held and all went well; they booked again for the next year while still at PSV. "Never a dull moment skippering PSV," *John chuckled.*[14]

Working with Haze was quite enjoyable over those twenty-five years John thought. Work and management standards were known, an incredible spirit was established, and never was there any plantation-like attitude existing on the island. Haze was responsible for that John believes. John and his wife of twenty-two years, Lusan, made their home on Bequia where he ran a consulting and accounting business. When the Corbett family stayed on the island during Haze's absences, their four children were taken over to the neighboring island in the workboat where they attended school at the Petite Martinique Roman Catholic Church School. John's eldest child, David, continues the sea faring life and is now staff captain with Norwegian Cruise Lines!

David Corbett on cannon at PSV

John remembers there was never a day that went by that was like any other; each day was very different. Once, about thirty pilot whales showed up around the island and on the reef, obviously in distress. They seemed to want to beach themselves although no one ever found out why this was occurring. Even though the resort was full at the time, John gathered staff together into various boats, trying to act as a guide to lead the obviously distraught whales out past Mopion Island. With everyone's heart-felt efforts, they were able to save about half of them.

Distressed whale at PSV

Although John enjoyed working close with many people on the island, he believes a very key person was Otnel Samuel:

"I had the pleasure of working with him for over two decades. When I was managing in Haze's absence, he functioned as my assistant manager. During my time there I believe that besides Otnel, no one other than Haze contributed more to the successful operation of PSV. Otnel was always a delight to work with and I cannot recall harsh

words ever. He always worked very hard to address the supply needs of PSV and a host of other issues. He did this with humour, tact and skill and earned the respect of all staff and management. I cannot recall a more competent manager in my years of work in the Caribbean or the USA."[15]

Although John enjoyed working with all the PSV staff, some particular people stand out in his mind. The first is Nora Mantz who was based in the offices in Cincinnati and was a major factor in PSV's success. She worked with travel agents and guests to get people to the island. "She had a gift of gab, a wonderful way of communicating, and was a very significant factor in getting PSV off the ground," John notes. Because of her 'personal charm' she was able to work out deals and bring in bookings. "She was delightful, always performing magnificently – a pivotal person."[16]

Boat captain Chester Belmar was also a very important person at PSV who John remembers as having a great personality; always greeting guests with smiles and hugs. Mattie Bethel had the distinct pleasure of being one of the first people many repeat guests would greet when they arrived on the island. They would rush to give her a big hug or ask her to lunch! John remarked that Goatie was and still is loved by all and that Eric Williams, the waiter and driver, had always been requested by so many guests to drive them around. He had a real personality and is terribly missed since his unfortunate passing a few years ago.

Eric playing cricket

Mr. Grant, who was a popular waiter, was nevertheless short and sweet with his responses, John fondly recalls. One evening the chef prepared a lovely 'white swan dessert' which was a frozen ice-cream treat that had been molded into the shape of a swan. "I remember a guest asking Mr. Grant to describe the 'white swan' to aid in his dessert choice. Mr. Grant quietly responded, "It's like a duck!"[17] John still chuckles every time he remembers this.

It may not have always been the staff, management, or any other person for that matter who 'stole the show'; the Richardson's dogs were the biggest draw for several repeat guests! Certain guests continued to grace the island with their presence mostly because of their love for the dogs John believes. Both John and Lusan remember one guest in particular always asked immediately upon her arrival for dog Hera, who stayed with her for the duration of her visit.

Postcard of PSV

During the early years John was working on the island, communication was lacking to say the least. There was only AM 2182 radio communication used except for telex. This limited communication with the outside world made arrivals, departures, and getting supplies often a challenge. And some 'mistakes' were made as well; John recalls Haze mentioning that during construction it was not until after twenty-two cottages were built, did the communication come through from Mr. Nichols saying that he wanted only twenty-one cottages built!

One major occurrence during the Corbett's time on the island was the Grenada Revolution of 1983. *Operation Urgent Fury* saw the United States invading Grenada, trying to rid the country of their then communist government.[18] During the time prior to the invasion, when People's Revolutionary Government leader of Grenada, Prime Minister Morris Bishop was in power, he gave a speech on Petite Martinique. John Corbett was present and thought Mr. Bishop made a good speech speaking about Abraham Lincoln and how he wanted his government to be like the United States, among other topics. General Hudson Austin was also with the People's Revolutionary Government of Grenada but now opposed Morris Bishop. In contrast, the speech he made was warlike. John was able to speak with Morris after the address by both parties, but sadly just a week later, John recalls hearing the news that Bishop, family members, and others had been brutally executed.

PSV was somewhat involved when the United States invaded Grenada, but in a peaceful way. The invasion happened in October and November 1983 while the resort was in its off-season time. There was however, regatta jump-up festivities being held each afternoon. A phone call was received from the United States military command to discuss possibly visiting PSV via helicopter. "We said they would be very welcome especially during the regatta and cleared the field in front of the restaurant," John recounts.[19] Troops would take a helicopter to the island where they could have a much needed respite from the on-going hostilities. They would arrive, put down their machine guns alongside the Petite Martinique band playing music and joined in the festivities, dancing and relaxing. These men were able to lay their problems to the side and unwind, calm down, and have a much needed break from the issues at hand, if only for a very short period of time. "We enjoyed the irony of the happy uniformed soldiers dancing to *Boots, Boots, Government Boots*, a humourous Barbados calypso critical of Barbados military spending. The visits were repeated during the several days of regatta festivities and we received thanks from the commanding officer of the United States forces."[20] The conflict soon ended and peace was restored on Grenada and on PSV.

Helicopter landing at PSV

Helicopter flying into PSV

Bob Law and Haze talking with the troops

Otnel 'joining' the Army!

Troops over PSV

Bob Law, Haze, and Marge Law with helicopter troops

Cole Beadon recalls "the young soldiers in the bar telling us that one of the biggest problems in Grenada was the large centipedes that were crawling into their sleeping bags at night and stinging them when they accidentally rolled over onto them – very similar to a scorpion sting and not at all pleasant!" [21]

Many a tall tale have been circulating throughout the islands over the years and a particular story John has heard many times, is when the People's Revolutionary Government apparently decided to introduce a customs officer on Petite Martinique. They had suspicions that smuggling operations existed there and therefore an official needed to be based on the island. According to John, a couple weeks after the gentleman's arrival, a real funeral was taking place for one of the residents who passed away. The officer was present in his official duty and all the inhabitants of the island were dressed in their Sunday best. But when they arrived at the cemetery, two graves had been dug rather than just one. When the official asked who the other grave was for, the people told him they built an extra one for him. That was apparently the end of the government trying to get customs officers on PM. Does this story sound familiar? A slightly different version from the one Mr. Nichols recounted!

Another tale of the Caribbean often heard concerns the reason the island was sold so inexpensively. It is told that the residents on Petite Martinique, even though they could have settled on the island themselves, did not want to develop it as they thought it to be haunted; it was not valued and it was 'scary land.' Everyone one is happy that tale does not hold true.

John and Lusan continue to live on the nearby island of Bequia and they are still having a great time sailing and participating in the regattas of the area as time allows. John remembers sailing in past Bequia Regattas with Sir James where every Easter the race is still run. The Corbett's time on Petit St. Vincent spanned many years and although each day was filled with enjoyment, excitement, and much hard work they recall those days with joy and laughter.

Corbett sailing with Sir James at the helm one Easter regatta

John Corbett on Freya *with Sir James at the helm*

John hard at work 1999

MARIN TANASE

In 2008, a person was needed to handle the mechanical operation of the island. Kenny Cordice was sent to find just the right person to fill this position. Kenny remembered a man in Trinidad who used to fix his boat so he went to the shipyard there and asked to see Marin Tanase. He explained the island's situation and its need for an engineer. At first, Marin was hesitant as he had happily worked in his current position at the Maritime Preservation shipyard for five years, but he flew to PSV to see what the job entailed. He was offered the position but agreed only on the condition he would work on a six month trial period before he and the management made up their minds. Not too long after, it was mentioned to him that there may be some major changes coming soon. [Unbeknownst to him, the island's ownership was going to change hands.]

Marin's background is rich with details about his life before joining the island. He is from Constanta, Romania where he grew up and attended college. After graduating, he moved on to engineering school where he specialized in diesel marine engines. To repay the government for the training he received, he was required to put in five years of service but he stayed for eleven years working for the Romanian fleet. He began as a cadet and then he was promoted to full engineer. He worked on very large ships weighing fifty-five thousand dwt [dead weight tons]. His many job duties included all aspects of the ships operation including maintenance of air conditioning units, generators, septic systems, lights, and navigation equipment just to name a few.

Soon after, Marin worked for Windjammer Barefoot Cruises out of Miami, beginning in 1994. There were seven different sailing ships but the line he worked on ran from West Palm Beach to Trinidad on a twenty-eight day cruise. It would stop at ports on fourteen different islands on its journey. The ship he sailed on was built in 1954 in Scotland as a gift for Queen Elizabeth. It was run by mechanical workings, not the electronic components more commonly found today. In 2005, the owner of the cruise lines promoted Marin to chief engineer, working in the Trinidad shipyard. He was responsible for maintaining the boats that were out of the water for inspection and repair.

It is a little different at PSV. Even though the work is challenging and demanding of his time every day, he works with much smaller equipment! Currently, one of his main duties is taking care of the watermaker . . . which is a project everyone especially appreciates! Marin was extremely impressed when he visited the island, especially how Doug Terman had designed, and subsequently Haze maintained and improved, the system it is today. He said "I knew Haze had an art and probably learned many of his skills from the Air Force . . . and that his first priority was water."[22] Marin realized Haze took his training from the military and had put it into civilian practice. He understood that although that there was no power at that time and no hot water, they were still able to design a system which instead brought in fresh water by natural flow.

The original watermakers could make twelve thousand gallons of water each and there were two of them. But this was not enough to keep up with the current demands. The island presently uses between fifteen thousand and nineteen thousand gallons of water every day on average. Marin suggested a proper watermaker should at least be able to handle twenty-four thousand gallons of water every day. Because the price was not going allow that to be possible, Marin is using both one new one which produces twenty thousand gallons and one old one. There is a gravity-fed water tank that is about half the way up Marni Hill which holds one hundred fifty thousand gallons. The other tanks that hold water are across from the laundry building. In total, four hundred fifty thousand gallons of water can be stored at any one time. Marin climbs the hill every morning to check the levels of water inside the tank and if there is a leak, he will know to check the restaurants, staff quarters, and beach restaurant first. Even a small leak could create major problems.

The septic treatment plant in place on the island is set up to clean the discharge and waste water from the cottages. It dumps into a saltwater tank under each cottage. The water is cleaned and discharged back into the sea.

A major change in 2011 came with the addition of a new power station. This is yet another area where Marin uses his expertise. A new generator replaces two old generators used for so many years on the island. It is used approximately ninety percent of the time, with the old units remaining as backups should the new one fail. The old generators are always on standby as well for exclusive use when the new one is being serviced. Every twenty days the current generator is shut down and service begins which includes changing the oil filter, cleaning the air filters, replacing the primary and secondary fuel filters, and adding new oil.

Another very important project that Marin participated in was the making of the new wine cellar. There was a rainwater tank, or cistern, which was attached to the pavilion; it had cracked over the years and was not being used [discussed in Chapter Four]. In 2011, an architect was sent to PSV to make recommendations for possible modifications throughout PSV. He walked around the island with Marin and discussed ideas and plans to implement possible changes. When they came to the restaurant, the architect wanted to extend the original bar gracing the pavilion. In order to do that however, it would have been necessary to cut into or remove a load-bearing wall in the pavilion. So that change never materialized. However, when they went onto the roof of the building, an idea struck Marin. He suggested that the old rainwater tank be turned into a wine cellar. He doesn't know exactly what inspired him at the moment but as they were up on the roof looking down into the round emptiness of the tank, the idea of wine popped into his mind!

This solved a lot of problems. Although the pitch on the roof would need to be altered and a door added, there would not be changes to the load-bearing wall and the rainwater tank could be repurposed into something useful and beautiful at the same time. The work was begun not too long after. The stones on the wall had to be carefully removed by hand, slowly chiseling out the mortar which was holding the stones in place. Once the opening was wide enough, a jackhammer was meticulously used to remove the remaining pieces. This process was very slow and the most difficult because Marin was afraid the vibration of the tool could make the entire structure collapse.

But all went well and the doorway was opened. The entire remodel took one week and was constructed by a number of people including Otnel and a specialist from Union. It is truly a remarkable wine cellar that, according to waiter Roland Regis, can hold approximately twenty-two hundred bottles of wine, Old and New World rosé and champagne, a variety of international and local age rums, and excellent Cuban cigars. The cellar temperature is set to 62°F. Wine tasting is often conducted in the wine cellar or restaurant and bar area.

Marin is a remarkable man who is very hard working. Although just joining the island a few years ago, his dedication to PSV is incredible. He works from sunrise to sunset with the rest of the crew making this resort a place to be proud of. He has weathered all sorts of emergencies including stalled watermakers, leaks, and power outages but he always continues on without complaint. The island is very appreciative of his years of service.

ALFRED GAYMES

An incredible account of achievement lays in the story of yet another one of the workers at PSV. Alfred Gaymes, also known as Slick, worked for the island for an incredible twenty-five years beginning at the young age of fifteen. He started working as a dishwasher in 1972, arriving from St. Vincent where his family lived. He came from a very impoverished background; his mother had thirteen children and his father left shortly after the children were born. Unfortunately, in those days there were no child support laws and mothers were sometimes left to fend for themselves and the children. Slick remembers some young women would have four or five children by the age of twenty four. Some St. Vincent women who ended up traveling would come back and teach their friends about valuing oneself and fertility control. Alfred did not get to complete all of his schooling because he had to stop and help his mother work in the fields. He did not have much opportunity where he was but that all changed when he began work at Petit St. Vincent.

He was quickly given the name of 'Slick' as he is Indian and had long hair at the time which he would 'slick' or comb back to give it a nice look. He said he hated it at the time but after awhile it grew on him and so it stuck, even to today. He enjoyed his job as dishwasher but knew he could do better and said Jennifer Richardson really pushed him to achieve. She would buy him books so he could teach himself whatever he wanted to learn. She would have him work with the chefs they brought in from other countries and he would sometimes get to train with them in the United States during the off-season. One time, he stayed in Boston for six months with a chef that had worked on PSV earlier that year. When Jennifer traveled, she would request her favorite recipes from the restaurants she and Haze dined in and ask Slick to try making them when they got back. "Jennifer was an unbelievable person," Slick said, "And Haze was great also at helping staff."[23]

Even though Slick came from a needy background, neither Haze nor Jennifer knew about this when he first arrived. After all, as Slick pointed out, that is not something a person would normally add to their resume when applying for a job. But when they did find out about his family situation, Haze set up a scholarship foundation to help pay for children of PSV employees to go to high school so they could achieve more and have more opportunities. Schooling was expensive as there were uniforms and school books to buy, as well as the cost of tuition. If the child was able to pass the test necessary to move from primary to secondary school, Haze would ask how much the cost would be for the family, and allow the employee to pay it back, usually taking a small amount out of their paychecks. Sometimes Slick remembers they would not even ask a family for remuneration.

Not only did Haze help out staff members on the island, but would also help organizations on neighboring islands. Many times people from the mainland or surrounding locations would ask for contributions to help with many different needs. Haze didn't seem to hesitate to write them a check, Slick remembers. At Christmas time, the Richardson's would make 'hampers' to give to St. Vincent homes. These were food baskets given to families so they would have nice meals around the holidays. Jennifer would go shopping for the staff as well and every worker would each receive two or three Christmas presents. She knew what kinds of things they wanted since they were like family to her and Haze. She would get them shirts or sneakers or if she knew they wanted a CD or a camera, she would make sure that was their gift. "Mr. Richardson and Jennifer had the greatest hearts of any persons I've ever met," Slick says warmly. [24]

During the time he was employed, Slick had run into many humorous situations. One in particular has stuck with him even to this day. He is a smaller built man, standing about 5'3, and with his long hair, it got him into a kind of uncomfortable predicament one evening. During the Wednesday night Jump Ups, Slick tells the story of how everyone was dancing and having a great time. He was enjoying a few dance moves himself when a man came up and started dancing next to him. Slick thought nothing of it since many guests were just having fun. But then the casualness escalated with the gentleman getting closer and closer, eventually putting his arm around Slick. This made him just a bit too uncomfortable, even if this guy was just being friendly. But the worse part came when the guy whispered in his ear that he would like to kiss Slick and take him home! This whole time, the man believed he was dancing with a long-haired Indian girl!

Slick was fit to be tied and got away very quickly. He was extremely upset and tremendously embarrassed. Haze came over right away and the two went and sat down somewhere away from all the 'action.' Slick remembers, as Mattie mentioned in an earlier chapter, that Haze always would start a conversation with a story and this night was no different. He started telling Slick that when he was in college, a very nice guy he had met would give him a ride to his classes. Every day this gentleman would pick him up at his apartment and drive him to the university. This went on for a year but finally one day, the guy asked Haze for a date! Haze did not have that particular preference and was confused and rather uncomfortable about the whole occurrence. In recounting this story, Haze wanted to make sure that Slick understood he was not alone in experiencing awkward and embarrassing situations in life and told him that these things will happen. This made Slick feel much better . . . until Haze just had to add 'but after all, you can't blame him, you are SO cute!'

His impression of Haze was one of admiration and amazement. "Mr. Richardson was a great man and one of the most loving human beings I've ever met. He was unbelievable. He listened to everyone's problems and helped everyone."[25] He does not believe Haze could ever say no to anyone and that ALL the staff always had great things to say about him. Back in those days when a boy would graduate from school, instead of presenting him with a car or something similar to what happens now, it was customary for the father to give his son a piece of land to start him off right. Therefore, Haze would always make sure if one of his employees wanted to build a house, he would loan them the money to be able to do this for his family. To show his support for the staff in other ways, he attended funerals of the workers so he could grieve with the family members. He would also make sure a proper burial was received at the cemetery. He sent condolence cards as well as 'just being there for support,' which meant a lot to everyone Slick remembers.

Slick will never forget when he told Haze he was going to get married. "He gave me many words of wisdom and then he said, "If you really love her, go for it!"[26] Mr. Richardson told me to plan a party and wanted me to invite friends who could stay in the cottages. "My friends could not believe they were staying on a private island. It was the greatest time in their lives!"[27] He decided he would cook for his own wedding which was quite a feat for sure. It was made a bit easier by PSV's meat supplier: he was from Boston and offered Slick anything he wanted since they were such good customers. By the time the wedding rolled around he was quite tired but pleased he had been the one to prepare the meal for his wedding guests.

His most prized recollection of Haze came after the marriage ceremony which Haze attended. Many people who knew Alfred offered a congratulatory toast but it was not until very last that Haze stood up "and that is when people really starting knowing the real me, the man from PSV," he reminisces.[28] He remembers Haze declaring, 'Everyone is talking about Alfred but his name is Slick and even though he looks small, no one knows what he has achieved. We put a stool under his feet to help him reach and clean the dishes and he became my personal chef. You guys don't know anything about him!' Everyone just burst out laughing and Slick said that speech made him feel like he was ten feet tall. "It was the nicest thing anybody had ever said about me . . . I never felt so important! Mr. Richardson even told the story about the guy who wanted to kiss me and take me home!"[29]

Mr. and Mrs. Richardson were both very dear to Slick. "I was always very close to Mr. Richardson and we would drink scotch together and play backgammon, sometimes all night. He was so competitive it was a fight to try to beat him." As for Jennifer, like the old saying goes, "Behind every great man is a strong woman," and Slick believes Jennifer was that woman. "She was the driving force and made the island run well."[30] He did have to tell her though when she first got to the island that she danced like a 'white lady', but he soon taught her how to dance West Indian style. "He was almost always one of my dance partners," Jennifer says. "He was a very significant member of our staff, our history . . . our family."[31]

Jennifer was always trying to help him improve especially as he moved up the ladder going from dishwasher, to salad chef, to breakfast chef, and ultimately to head chef. [Before him, every other cook was brought in on work visas from other countries to stay for about one year.] He was so proud to be bestowed the great honor of being the very first local person to attain this title. A couple months after this accomplishment, he decided to take a holiday to St. Vincent. When he got on the ferry, he realized Sir James was aboard! Slick decided to go up and talk to him about the progress of the country and to give him his good wishes. After a lengthy conversation, Sir James asked Slick who he was. When he explained he was the chef at PSV, Slick remembers Sir James saying he had thought it was a white guy and offered him his congratulations. This made Slick very happy. Even the guests were impressed and had similar comments. They always thought a chef from another country or a 'white guy from a culinary institute' was cooking because they loved the presentation and the quality of the dishes.

Haze was the genius behind the scenes in the ordering of the foods, especially the meats which allowed such excellence to be produced. He decided to get his meat from Boston and had to devise a plan to accomplish this task. Haze arranged to have the order trucked from Boston to New York, put on an American Airlines flight to Barbados, and then repacked into smaller containers for its flight to Union. The final step was to have it loaded onto the boat from Union to PSV. This entire trip was accomplished all in one day! Slick said he would tell everyone who asked how this process worked and their next question would always be, 'Who is the brilliant genius who thought of this?!' Haze would even give the guests the butcher's contact number in Boston and many ended up buying meat from him, even if they had to have it shipped to their own homes.

Another story Slick remembers has to do with the fishermen who resided on the island in the 1970s. Although St. Vincent and the Grenadines does not allow for any private beaches on any of its islands, it does restrict free access to the high water mark of the sea. Past that, a person could be considered trespassing. There was not much industry other than fishing and tourism in the Grenadines and therefore many men from Bequia would fish or dive for conch and lobsters. Because they wanted to remain in the area as it was too far to travel back and forth to Bequia each day, the fishermen decided to camp on the beaches of Petit St. Vincent.

According to William Pringle, an avid sailor in the area better known in the Grenadines as Billy Bones, "the local fishermen had the right to pull their boats up to the high watermark"[32] which moves with tide on any shore, to conduct their business, with vegetation defining public area. This included sleeping, cooking, and corning [or salt drying] their fish. This was the way of life apparently for hundreds of years on Mayreau, Union Island, Canouan, and also on Petit St. Vincent. "There was a nice two-acre salt pan that flooded at high water where they corned their fish," recalls Billy.[33] Evidently, the process of corning fish released some pungent odors that were not pleasant to PSV management or the visiting guests.

Also, their campsites on the beaches created a number of problems for the island since it was a private resort and sometimes there were thirty to forty fishermen living there at any one time. There were some complaints from the guests who were trying to have leisurely strolls along the beaches. The fishermen did not in any way purposely cause problems but some people were intimidated by the sheer number of men camping. Although these were problems, the island did benefit from their work. PSV bought their catch every day which allowed fresh fish to be offered to the guests; 'from the sea into the pot' they called it. So there was a dilemma that had to be worked out; a plan appropriate for PSV that was also suitable to the fishermen. According to Slick, Haze did not have the authority to put any of the men off the beaches so he decided to form a delegation to talk to the authorities of the government.

He chose eight or nine key workers including the head bartender, waiter, and chef, who went to St. Vincent to talk to Sir James. Slick was among them. When they arrived, Slick remembers it was not the best of situations. Slick thought that Sir James was in quite a predicament to come up with a solution. On one hand, he wanted to support the fishermen as they were residents of the Grenadines. On the other hand, he wanted to support one of his country's other islands. According to Slick, Mr. Richardson's point of view was he wanted to support his guests and his employees, but realized the fishermen deserved to continue making a living. They went back and forth, trying to come up with a workable compromise.

Then "Mr. Richardson came up with a brilliant idea," Slick recalls.[34] He told Sir James he would build a proper house for the fishermen on the island. Slick noted how pleased the Prime Minister was and mentioned that it was the most interesting thing he had ever heard. Haze wanted to improve the situation for everyone but had a couple of conditions which must be met: There would be no more than fifteen to twenty men allowed to live on the island at any one time, they must keep the quarters clean, and although they were free to be at the bar, if there were any problems, such as intoxication, the police would be notified immediately. The terms were agreed upon and the construction began. It was a two-story home that had proper rooms and bathrooms, just as nice as the staff quarters, Slick thought. It all worked out with PSV continuing to buy fish from them and it appears that there were no issues with the fishermen from then on.

In all the years of service Alfred Gaymes put into the island, and all the good and not so good happenings in that time, if he had to relive his life again, he would chose to come back to PSV because it made him so happy. In his opinion, Haze and Jennifer Richardson made it a paradise away from home for the workers. We were all treated with respect and Mr. Richardson was like a father to many, which Slick believes is a great tribute and legacy Haze left behind.

ROLAND REGIS

Roland Regis began working at PSV as a waiter in 1997 and is still employed there today! He is one of the guests' favorites and is repeatedly asked for when they return. Many consider him quite a good friend and share numerous hours talking and catching up. Roland remembers his beginnings on Petit St. Vincent when his mom was 'able to pull a few strings' and get him a personal interview with Mr. Richardson. A PSV boat was sent to pick him up in Union where he lived and then he and Haze had a nice, productive meeting, Roland remembers. He was asked to start work that Friday, December 3rd. Roland has a very high level of respect for Mr. Richardson, ever since day one. He felt the island was a labor of love for Haze. He was not so much a people person but loved talking to the locals. "He was the boss but I didn't feel he was on top. He was the one who always pushed you to be there and to be your best," Roland said of Haze.[35] But he also did not care for loafers. He would offer letters of warning to the staff members and therefore there were seldom any people who were fired. "On a scale of one to ten, I would say Mr. Richardson was a three on firing." Roland remembers him always saying, "You know who you have; you don't know who you're going to get."[36]

Feeling very close to Haze was easy for most of the staff, especially Roland. "He would tap me on the back and say, "You are a star Roland" and that meant a lot to him because Haze always prefaced that comment by saying "You can teach a person the trick of the trade but you can't teach him to be a star."[37] Roland had some mighty fine people to emulate like Elvis King, nicknamed Boopsie. He was a very popular waiter and his great service and fine character inspired Roland to be as good as him, or maybe even better if he worked very hard! Roland remembers Haze saying, "Until you get to Elvis, you are not there!"[38] Wow, what big shoes to fill!

Many different assistant managers have graced PSV but none of them worked out for very long. According to Roland, none provided the personal attention Mr. Richardson did but all were able to leave their mark, for good or bad, in one way or another on the island. "One particular manager," Roland recalls, "used to call Haze 'Boss' which he didn't like. He much preferred being called Skipper."[39]

During his free time, he enjoyed working on computers. Haze therefore decided to give him a US$500 loan to be able to buy one for his personal use. Roland was very excited and wanted to help bring the new technology to the island workers as well. He was instrumental in setting up a computer station in the engineering room which was adjacent to the workshop. With the help of Kenny Cordice, resident computer expert, there were four computers available with internet service and everyone was able to get a PSV email address. Even Cedric the carpenter got involved and built stools for the room! Roland soon started a 'computer club' and Kenny offered to give lessons to the staff. He even installed Roland's new computer at his home when it arrived. Roland remembers a funny time when Haze was frustrated with needing a coupling part for a machine but was unable to describe what he wanted to his contact in Miami. Roland said, "Just Google it." Haze couldn't believe it was that easy. And this was the start of him being inspired to get the staff more computers and training. "It was a big time back then," Roland laughs.[40]

Getting along with Haze was easy for Roland however he feels that many staff members were sometimes intimidated by him. Therefore, they never sent any new guy to wait on Mr. Richardson for room service, in his house, or in the bar! Out of respect for him, none of the staff called him by his first name; it was always Mr. Richardson or Skipper.

During the off-season, the staff would often have a hard time making ends meet with family responsibilities so they were given a bonus which arrived about three weeks after their last day in September. Haze would personally assess each staff member on their performance even if he heard comments from other employees or guests. Included with the bonus, there would be a letter from him outlining the year they had, along with improvements the island was making during the closed season, and each person's return date. 'Get plenty of rest!' Haze always said. Roland thought it was very kind and thoughtful of Haze to not just send a check but to include a personalized letter as well.

Unfortunately, there was a problem with bonuses at the end of the first year new management took over according to Roland and other staff members. The bonuses given each year were from a mandatory gratuity included with each guest's bill. But instead of receiving a check once they got home, each department was called to meet with the accountant at the time and was given an envelope containing cash. Roland was handed EC$300 for the entire year's work rather than his usual average of around EC$7,000 and the others were handed similar amounts, some much less. According to the staff, they never received either a satisfactory reason or the rest of their due monies. Because of this, it was voted on by the employees to have their bonuses given to them monthly rather than waiting until the end of the season.

When Phil Stephenson took over the island, Roland however felt they hit it off pretty well although things were definitely different. Everyone was family when Haze was around but when new management came in rather than an owner being present, everything was disoriented. Roland told Phil, "I think the ship runs better when you are here."[41] But Mr. Stephenson does not live on the island so changes are eminent. Haze blended well into the culture of the islands and could speak their 'language.' He spoke words they spoke and was able to use them at the appropriate times. The management team now is different and Roland notices quite a difference from Haze's style. Many staff members are having the same difficulty transitioning into this type of administration Roland feels. A short time of this is to be expected when major changes happen as long as morale is not lost forever; a difficult balance for sure.

Roland has been extraordinarily happy with his time at PSV during these past sixteen years. He has dedicated a very large portion of his life to the island and received in return numerous awards such as an *Outstanding Staff Award* and is a two-time receiver of the *Guest Award*. Although there are many highlights during his employment, two stand out especially in his mind. The first one was when he was given a trip on a cruise ship in 2003 to Cozumel when he won the outstanding staff award. He, along with Goatie, Mattie, Delia, and three others had an unbelievable time! The second highlight of his career so far was him being selected by Mr. Stephenson to attend a wine training course in Miami which he enjoyed very much.

The island has forever been improved by Roland's winning smile and kind words. He continues to be many guests' favorite server and he receives that praise humbly. He is more interested in making sure everyone has an enjoyable time at their home away from home and that is very evident in the smiles on their faces when he is present. He feels his transformation during the time he has worked at the resort has been one of great personal growth and has afforded him incredible experiences that otherwise would not have been possible without Petit St. Vincent. He is forever grateful for his journey and the island feels the same to have such a wonderful employee as Roland Regis.

Roland sporting uniform from early 2000s

ROLAND RICHARDSON

The island has had the distinct pleasure of having the works of world famous artist Roland Richardson displayed on their walls since 2012. He is considered a Plein Air Impressionist painter whose creations have graced the Caribbean for over forty years. Through his art he communicates the light, colors, and images found throughout the islands, which is a tribute to his French Caribbean heritage. His masterpieces are always created in the field from something living to inspire him and to honor nature. He transforms landscapes onto canvas so that all can enjoy the beauty of the area even when they are not part of it.

Phil Stephenson has been very fortunate to have met Mr. Roland Richardson and is proud to showcase his exquisite artwork on the island.

Roland Richardson's painting of a flamboyant tree hanging in the PSV pavilion

Painting Lesson
Roland Richardson and Jack Stephenson

CHAPTER TEN

THE ESSENCE OF PSV - THE GUESTS

The Nichols' dream of a resort materialized in 1967 and as Mr. Nichols said: "It is the concept of Petit St. Vincent that it be a quiet family-type resort intended for those who want to get away from the telephone and have a quiet vacation."[1] Little did he know at that time the island would become a haven for so many for decades to come . . . not just for new visitors but for returning guests who chose PSV as their home away from home year after year. They are the key to a successful resort and PSV's guests have proven to be the finest in the world.

The Nichols' daughters continue to visit the island when they can. Beth has fond memories of certain staff members and searches them out as soon as they arrive, as do most other repeat guests. Her favorite friends include Mattie, Goatie, Otnel, Roland, Slick, Liverpool, and Joseph. When she was young and visited the island with father and mother, she remembers a very young man named Chester Belmar paddling up in a dinghy, trying to sell fish to her family. This was when he was not quite old enough to be able to work on the island nor was he able to get his license to operate a large boat. However, he later became one of the most important boat captains on the island and contributed immensely to the smooth operation of the dock house facilities.

Chester and Goatie

She also remembers in later years her children would accompany her to the island and Joseph, the sous chef, captured her family's heart by bringing them a very large chocolate bar from the kitchen! It is little gestures such as these that make each visit special and when the guests come back, it is as if they never left in the first place. Time has a way of standing still on Petit St. Vincent and it is quite magical.

THE BLISSETTS

Malcolm and LeClaire Blissett [Lee for short] have been coming to the island as regulars for over twenty-six years! And they did not spend just one week . . . they were serious about their 'home away from home' as they called it. They started out staying for two week periods and that quickly progressed to spending six weeks at a time on PSV for six years straight! For eighteen years, they stayed in the same cottage, #3. Their big dream was to come down from mid January to the end of

April so they could miss the harsh Michigan winters where their other home is located.

Throughout this entire time, Malcolm noted that only a few changes have happened on PSV. There was a driveway added so the mokes could pull up close to their cottage and a covered patio was built which made a nice place to enjoy breakfast or a cup of tea.

Like many other guests, they were looking for a place that would respect their privacy but where they could connect with people if they wanted to. They found the island offered all of this and more but ultimately, they chose to return because of the friendly staff. The Blissetts consider them to be the soul of PSV and are responsible for the feeling of 'coming home' which guests get when they arrive. Malcolm and Lee were always warmly greeted on the dock and the staff made sure, as with every guest, their needs were continually met. But the staff went a little extra out of their way for the Blissetts. One night when they didn't show up for dinner at their usual time, Goatie got so worried that he went down to their cottage to check on them. Malcolm remembers as if it were yesterday how they heard a "tap tap tapping" on the window. It was Goatie just making sure they were ok! This special attention meant a lot to Malcolm and Lee.

As repeat guests, they were excited to meet other guests who also came back time and time again. After a while, they became very close friends with two other couples from the United States. And in honor of their friendship, the six of them decided to form a club. They named it the 'PSVI' [the Petit St. Vincent Six or the PSV Six]! This club and their repeated visits to the island became a major focal point in all of the couples' lives and they looked forward to seeing the same people each year. The official PSVI members were Malcolm, who was named the 'Grand Poobah,' Lee who was called 'Her Highness' since she was married to Malcolm, a lady named Kathleen who decided she would be a 'Princess,' and her husband Hal, who was of course referred to as 'Earl.' That left Deb who is always a 'Lady', and her husband Les ranking as 'Duke!' Malcolm even had shirts, hats, crests, and special flags made bearing the emblem of the PSVI. The staff graciously went the extra mile by designing and installing special flag poles on which the PSVI flags would be raised when the group arrived to the island!

The PSVI

PSVI crest and Malcolm Blissett holds the flag he designed

Because they stayed at the resort for so long each year, they have many stories of some of the antics that went on. One year they remember some guests started getting very possessive of the beach hammocks. According to Malcolm, a long-time guest from the United States went running to the office to report that a Russian guest had taken her spot and the intruder must be removed immediately! That guest went on to boast, "I've taken the island back for America."[2] Another time that same year, guests began asking staff members to come very early in the morning to take their towels and bags down to 'their hammock' so that it would be available when they wanted it. But they would go off to have breakfast or whatever, taking some time getting down to the beach. This angered the other guests who actually showed up at the hammock and were ready to relax. It was a big deal at the time and everyone was getting in a huff about their supposed rights to each hammock! As the story goes, it apparently got so bad that an English boxer, infuriated that his spot was taken, punched the interloper right in the face![3]

When Haze learned of these escalating troubles, he decided something had to be done. Although he was sympathetic to the guests' desires, things were getting out of hand. He finally realized why each guest kept thinking a particular hammock was theirs; it was because the shelters had numbers on them. So if they were staying in cottage #2, for example, they believed they 'owned' the #2 hut! Once that was figured out, the problem was easily rectified: the shelters were adorned with beautiful new plaques that had letters on them rather than numbers! And Haze instructed the staff they could no longer offer guests bag service to the beach hammocks . . . they would just have to go claim whichever one they wanted themselves . . . on a first-come, first-served basis!

Malcolm sheepishly admits he was not without guilt during the hammock holding humiliation days of PSV. He points out that he and Lee also had a favorite place on the beach. Wanting to make sure they always got that particular spot, he admits he often went down to the hammocks in the early morning hours. "I would put a book, a towel, and a cheap pair of sunglasses onto the chairs around the hut so it looked as though someone had been there."[4] One day it got so bad he confesses, that when he realized one of the two chairs normally at 'his hammock' was missing, he sneaked across to the other neighboring hammock to take a lounge chair from that one. To make matters even worse, he declared, "I decided to carry it back through the water so that the chair's tracks wouldn't be seen in the sand and there would be no trace of the crime committed."[5] He laughed when he remembered how silly he felt "spending all this money coming to a private island and acting like a spoiled kid."[6]

One of the lovely beach huts

The Blissetts always thought highly of Haze saying they really enjoyed the time they were able to spend with him and they particularly liked the way he treated staff members. Malcolm pointed out that not only was Haze very creative but that he was also very insistent on honoring guests' privacy. "Haze told me, 'this is the drill: privacy and no tipping.' "[7] Haze did not want the workers to help someone or give them special service solely on the hope they might get a tip. Instead, there was a fund set up at the time for the staff to which guests could contribute if they chose.

Sadly in 2010, Lee Blissett passed away. Malcolm chose to return in 2011 and when he did, he threw a cocktail party in her honor and invited all the staff and guests on the island. The workers planted an orange tree for her to be remembered by and Malcolm had a special plaque made which was put up on the wall outside cottage #3. It exists as a reminder of the many years of enjoyment shared on this magnificent island between Malcolm and the love of his life.

Memorial plaque for Lee Blissett

Another couple has been regular guests since 1970! Even though they wish to remain anonymous, they shared many stories about their time on PSV. They first came to the island specifically to avoid the gambling, cruise ships, and large hotels that were 'ruining the old Caribbean.' They sought solitude and in cottage #1 they found it!

In 1970, there were a few differences from today they remember. For example, mokes [discussed in a later chapter] did not exist on the island but that did not stop meals from being delivered . . . they just arrived by bicycle. Also, each cottage had folding chairs that guests would carry down to the beach if they wanted something to sit on. They recalled that year the staff, including the chef and other key people, walked out over some kind of disagreement. Therefore, luncheons and dinners consisted of "frozen Sara Lee things and Campbell's soup!" they recall.[8]

Similarities in the resort as compared to today include the beach cookouts that happened once a week. Tables were set out on the sand and torches were used for lighting. Lunches could also be ordered and delivered either to the beach or to their cottage. Although there was no spa on the island as there is now, they were able to enjoy massages from a lady named Genevieve who lived on Petite Martinique. She would simply come across in her dinghy bringing her massage table. They would request her services on a signup sheet in the office.

They recall every Thursday night there was a cocktail party at the Richardson's house. They liked this opportunity to talk to Haze and mingle with other guests. The couple fondly speak of him, remembering how kindly he treated the workers and how he was regarded as a sort of father figure. They saw how Haze loved the local people and tried to help them solve problems if they did something wrong rather than just firing them.

So why do they continue to this day to chose PSV as their home away from home? For many reasons but their number one motivation is the quality and caring nature of the staff. Also, they feel the island is a "bubble in the region . . . a rare gem."[9] It is unique, where the lack of intrusion is held in high regards. It is simplistic but sophisticated, where less is more, and is natural but manicured. This year of 2014 marked their 44th anniversary of coming to the island.

THE MAXWELLS

Ann and David Maxwell are long time repeat guests as well. They have 'come home' every year beginning in 1991. Ann remembers how, although they stayed in different cottages on the Atlantic side of the island throughout the years, they liked #6, #7, or #8 for the particular privacy of that end of the beach. Their most recent favorite is cottage #19 because it has its own beach with a hammock and terrace overlooking the harbor. Loyalty such as theirs is wonderful but coming home to PSV is something that does not require a lot of thought; it's just natural to be drawn to this island paradise for so many unique reasons and the Maxwells found that out year after year.

During their first trip to PSV, Ann remembers asking Haze a question about the vegetable garden and he ended up giving them a private tour. "He talked with great enthusiasm about all the things he was able to grow for the kitchen"[10] and also loved talking in detail about other practical aspects of running PSV as well. They were always happy to be invited to dine with the Richardson's and did so often during their visits. "That was always a good evening and Haze would entertain us with lots of amusing anecdotes," Ann fondly mentioned.[11]

Many other warm recollections of Haze fill the Maxwell's hearts. "One of my abiding memories was of him sitting or standing at the bar in the evening with his dry martini talking with the guests. He was a very jovial and welcoming host."[12] One particular evening they were dining at the Richardson's home and Ann realized, "He was probably at his best and most entertaining in a small intimate group. He was extremely good at recounting incidents on the island and on his travels."[13]

Over the years, she and David became aware of the fact that he was "hugely respected and loved by all the staff and that he regarded his staff as part of his own extended family. He had also earned the respect of local people on neighbouring islands. I remember him being delighted when a guy from Petite Martinique came over and made him a gift . . . of pigeon peas from his garden. Little things like that meant a great deal to him," Ann confided.[14]

Not only was Haze essential to the island, but Ann and David remember how important the dogs of PSV were to island life as well. They would regularly visit guests in their cottages and run up and down the beaches. Their special friend was Zeus. "He would demand that we throw stones over and over again for him to retrieve. He and the four puppies, Daisy, Dotty, Ducky, and Mr. Green, would appear at our cottage when our breakfast was delivered and look longingly at the food on our plates."[15]

David Maxwell held his 70th birthday celebration on Petit St. Vincent in 2008 with Ann, their children, and grandchildren. A great time was had by all. David and Ann will always hold a special place in their heart for the island and most value PSV as being "unique, laid back, and very private."[16] Sadly, David passed away in September 2012 but his memory lives on; a tree was planted as a remembrance to all who knew and loved him.

THE KENDALLS AND THE REEDS

Not only do couples come to enjoy the island but families as well. Jeremy and Lynn Kendall have been coming to PSV for twenty-two years along with their children and grandchildren. Daughter and son-in-law Penny Kendall-Reed and Stephen Reed [fondly called Steve] have also enjoyed the island for many years making it their 'home together' with the Kendalls, away from their separate homes in Canada. "It really is a family favorite vacation destination," explained Lynn.[17] The elder Kendalls even celebrated their 50th wedding anniversary here. Not only are they guests but have also become an integral part of the island through their participation in a new scholarship fund that was put into place in 2014. The board of the Scholarship Education Fund is presently made up of Philip Stephenson, Jeremy Kendall, and Noel Victory [Goatie].

While visiting the island, the Kendalls and Reeds regularly visit schools in the area to see what the children need in the way of supplies. They take a list of their 'must haves and wants,' package up the items in Canada, and ship these items down to the anxiously awaiting kids! "Jeremy and I support three schools each year in PM and Carriacou and visit the schools when we are in PSV. When you are lucky enough to be able to spend time at PSV we believe it is important to give back in any way we can," says Lynn.[18]

In 2011, Phil and Lynn ended up discussing fresh ideas he had for renovating the island. She shared her opinions with him at that time and continued after their initial meeting to discuss the topic by email. He wanted her specific views about adding televisions and modern conveniences other resorts have for their guests' use. Lynn pointed out that "this island is a haven for people seeking off-line luxury."[19] According to Phil, Lynn was instrumental in helping the wishes of many guests be known so he could decide what may or may not be the best plan to move forward.

But it's not just about relaxation and quiet on the island; the Kendalls and Reeds enjoy a lot of excitement as well. Because they are repeat guests, they know boat captain Jeff Stevens very well and usually take a day's charter every year. Many memories have been made on these voyages but one story in particular stands out Steve recalls:

"As it happened, one charter fell on April Fools' Day [much celebrated in the Caribbean!] so we naturally came up with a plan to lay a prank on him [Jeff]. We composed an elaborate letter with official-appearing letterhead and signatures and had it hand-delivered to Jeff by Otnel, who was in on the joke. The letter claimed that Jeff had not complied with local fishing laws and was due in court in Grenada within a couple of days. Preoccupied with the day's trip, Jeff tucked the letter into his pocket to read later." [20]

As the day continued, everyone boarded the boat and had a wonderful sail to Sandy Island off Carriacou [which is in Grenadian waters] to go fishing. In the meantime, Jeff, still unaware that a trick was being played on him had decided to play a trick on Steve with the Kendall's help. In past years, their on-board fishing contests emerged with Steve as the most recent winner. So while Steve wasn't looking, Jeff placed a very small frozen fish onto an enormously large hook and cast his line. Everyone else joined in the fishing fun and the contest began. Steve remembers, "I was called urgently back to the stern with the announcement of a bite on my line and with mounting excitement I started to reel in. Of course, as the lure cleared the water and the sad and pathetically small fish appeared, he faced the derision of the crew! Much hilarity ensued as the prank was finally revealed."[21]

In the meantime, Jeff had recalled the letter in his pocket. "As they headed out for some snorkeling, he opened it and read the contents with gradually increasing panic."[22] The letter was so official looking and so convincing, Jeff never suspected a thing. When the Kendalls returned to the boat after about an hour, they let him in on the fact that it was all just a joke and ended his suffering. It was especially funny that both Jeff and the Kendall family were playing jokes on each other but not everyone knew it was happening! "Needless to say, I think both parties are hesitant about chartering on April Fools' Day in the future," laughed Steve.[23]

Steve Reed loves the island so much he decided to write a beautiful piece of poetry about PSV. It can be found on page 293. It is very befitting for such a beautiful island and PSV is appreciative of the extraordinary sentiments this offers. It speaks volumes of just how much the island is loved by all.

CHAPTER ELEVEN

ISLAND VEGETATION AND TRANSPORTATION

The island of Petit St. Vincent was sought out by Mr. Nichols because "it lay in the middle of the Grenadines, which to us, were the most attractive string of the islands in the Antilles;"[1] One reason he was able to say this was because of the island's vegetation. As was mentioned in an earlier chapter, the island was originally used for the production of Sea Island cotton and there remains a few of these plants even today. Although the island has changed over the years since the Nichols first saw it, Petit St. Vincent continues only to get more beautiful. Coconut palms were brought in by Johnny Coconut, Haze planted large amounts of palm trees as well, and throughout the years, new landscaping has been incorporated into the already lush environment. And with the addition of 'transportation', guests are able to move around more easily, enjoying the unbelievable scenery and vegetation.

Cotton from original island plant

The current head gardener, Roy Doyle, began working for the island in 1992. Roy had been employed for three years with the British High Commission on the island of St. Vincent as a gardener, sowing seeds and taking care of plants. However, when the BHC closed, Roy came to PSV. In 1998, he was offered a supervisor position by Haze and currently he oversees twelve staff members. Roy says a challenge of working on Petit St. Vincent has always been the fact that these soils are different from the other Caribbean islands. He sometimes finds five to six foot deep layers of clay under the surface which take a long time for water to soak into. This makes for a demanding job of having to 'baby the plants' in order for them to get a good hold and continue to grow properly.

Each plant or tree has a specific purpose and contributes in immeasurable ways to the island's natural world. Together they provide shelter from storms, offer homes for birds and insects, and supply food for animals as well as humans. Linda Leroy worked on PSV from its beginning in 1966 until 1969. Although she was the first payroll person on the island for the over two hundred men working in 1967, she was also instrumental in bringing many trees and shrubs from the botanic gardens in Grenada on the *Starlight V* mail boat which she planted around the island.[2] Upon a recent visit to PSV, she proudly saw that many have grown to maturity today.

Although there are many varieties and types of trees, the most abundant on the island are red and white cedar, with corkscrew and sea grape numbering a close second. The red cedar tree is of course known for its beautiful wood and aromatic fragrance. The white cedar tree is used very often for its fine wood as well and has a beautiful flower which blooms before the tree starts to leaf out.

The sea grape, *Coccoloba uvifera*, is an evergreen shrub that has a leathery-type leaf. It grows white flowers that bloom on a spike about 10" tall. Workers often use the foliage as decorations for events or as presentations for foods on dinner buffets. The sea grape produces a fruit which grow in bunches and is where the name originates from. This fruit is red in color and edible by both people and animals. Many of the pigeons on the island enjoy this fruit and its seeds. Roy says the people on the island are constantly fighting with the birds to see who can get to the fruit first! The seeds can also be used to make a natural red dye and is one plant the Arawaks and Caribs possibly used to color their rugs.

Sylvester Roberts preparing for wedding using corkscrew leaves

Flamboyant leaves grace the aisle

The sea grape grows well on the sandy beaches because of its high tolerance for salt and equally as well in the soil around the island and is therefore quite common. It has been used for medicinal purposes for hundreds of years throughout the West Indies. People use the boiled down bark to produce a gum which was supposed to treat throat ailments and it was believed that dysentery could be cured from the roots of the plant and therefore was an especially common herbal remedy during the cholera epidemics in the mid 1800s.[3] Because of its extraordinary beauty, the Blissett's used the sea grape to elegantly grace the design of their logo for the PSVI and is shown in a previous chapter on the flag he created.

Mahogany trees, *Swietenia mahogany*, were planted in 1994 around PSV and grow very fast. They are beautiful trees known of course for their spectacular wood. Because of the climate in the West Indies and its continual growth, it is known for its very large size and does not contain defined growth rings such as other trees do. The color of the wood gets richer as the tree matures and seems completely resistant to rot and decay. Its wood is considered one of the highest grades of lumber available.[4]

The Neem tree, *Azadirchta indica*, is another indigenous tree on Petit St. Vincent. It is actually in the mahogany family. The oil from its seeds is good for repelling mosquitoes; the seeds are dried and then the oil extracted and rubbed on the skin. The older inhabitants of the island put the leaves in boiling water and make a tea which they say is 'good for pressure.'

Coconut palms, *Cocos nucifera*, are widely spread around the island and were planted here by John Caldwell of Palm Island, as previously mentioned in another chapter. The tree was named 'coco' by the Spanish which means 'monkey face' because of its three indentions on the bottom of the seed. It can bloom as many as thirteen times per year and has one of the largest seeds of any tree in the world.

The coconut palm is actually not a nut, as the name might lead you to believe, but is a drupe, which is a kind of stone fruit like apricots, mangos, and plums.

It is a very versatile tree as every part of it can be used for one purpose or another. On the beaches and outside the cottages, the branches are used in the construction of huts. Many of these huts are also located at various other locations around the island; at the dock where guests coming to the island are greeted, just outside of the pavilion, and at the dining room and beach bar which serve as covers for the dining tables. Many guests enjoy reading a book or having a cup of coffee underneath one of these beautiful shelters.

Ocarol Nero, Gregory Baptiste (known as my friend), Roy Doyle, and Carlton Baptiste constructing a new beach cottage hut.

The trunk of the coconut tree is used for lumber and the seed fiber, or coir, from the husks is tied together with string and used to make brooms and brushes. The fruit is widely used on the island as well; the flesh is eaten freshly cut or is dried and used in cakes, pies or as a garnish. The restaurant offers a satisfying drink of coconut water from a freshly-picked coconut! The top is cut off and a straw is inserted so you can quench your thirst with its pure super-hydrating liquid! Another use around the world is using its oil; the dried coconut is pressed and the oil is extracted. It is used extensively in cooking as frying oil, put in soaps and in some cosmetics. It is high in saturated fat and can be stored for over two years because it will not easily become rancid.

Coconut Palm tree and staff member 'Coco' holding a freshly opened coconut

For hundreds of years, coconut water has been known for its many uses. It is said in the West Indies that when proper sterilization techniques could not be found, coconut water was poured over the infected area before surgery. Apparently the water from these young coconuts was thought to be very hygienic. During World War II, many dehydrated Japanese and British patients were given the water intravenously when the supply of saline solution was running low. This practice is not in use today as coconut water can cause elevated blood potassium and calcium levels.[5]

An indigenous tree to the island is the manchineel tree, *Hippomane mancinella*. It has been referred to as the Caribbean king of killer trees! As mentioned in an earlier chapter, this tree was used by the Caribs in various ways to poison the enemy. The Spanish referred to it as 'manzanilla de la muerta' [translated as 'apple of death'] because its fruit resembles apples and is considered one of the most poisonous trees in the world. Americans chose this particular tree to grace a movie in 1965 called *Wind Across the Everglades*. Actor Burl Ives ties his enemy to this tree and as you can imagine, the man was poisoned to death. And according to medical literature, the ingestion of the trees' leaves or sap can cause gastroenteritis with bleeding, bacterial superinfections, and shock.[6] And Ponce de Leon's death was attributed to the manchineel tree's poison, according to historical reviews.[7]

Manchineel tree sign on PSV

Roy unfortunately found out firsthand how dangerous it can actually be. During a construction project, water was taken out of one of the stagnant ponds to make concrete. The water was contaminated by rotting leaves from a manchineel tree and some accidentally splashed up on his face and ran into his eyes. Although it didn't start burning until about seven o'clock that evening, he found he was in severe pain. "It felt like a stone was inside," he said of the feeling if he closed his eyes.[8] Many people offered Roy their alleged cures such as flushing his eyes with sugar water or milk. He even tried swimming in the sea that night when he was in pain to see if the salt water would lessen his symptoms. None of these so-called remedies worked. He realized only time was going to help him.

In order to prevent poisonings in the future, he wears eye protection. He also said that since many people have a sensitivity to even the vapors, men will rub rum on their skin so that the fumes of the manchineel tree cannot penetrate to it. Don't let this tree scare you away however; although poisonous, it beautifully adorns the island and is not a cause for concern. You will be able to identify the tree right away as a manchineel because there are signs warning people not to get close to them!

On a much lighter note, a favorite tree of many on the island is the flamboyant tree, *Delonix regia*, or otherwise known as the Royal Poinciana tree. There is a very large painting of one in the bar area that was painted by Roland Richardson. Many of these trees adorn the island in various places. They have beautiful orange and red flowers that are glorious to see when they are blooming in April and May. Because of this display, it was given the name flamboyant for its show of flowers as well as being called the flame tree. The flowers are used as various decorations in the restaurants as well for weddings.

Flamboyant tree near the pavilion

Another PSV tree is the tamarind, *Tamarindus indica*. It is a native tree with many uses. It is mostly found on the sea front rather than on the mountain side since it can resist wind-born salts. As the tree gets older, the bark gets heavier. The wood has a beautiful rich red color and can be used to make furniture as well as flooring material. Its fruit, which grows as a pod, is made into candy, called tambran balls, and is added in various cuisines around the world. Fedelin, Lily's daughter-in-law, still makes delicious tambran balls today. Other uses are for medicinal purposes where it is believed that a poultice made with the tamarind plant will take down a high fever if placed on the ailing person's forehead.

There are hundreds of different kinds of plants and trees on the island and Roy speaks about all of them with great enthusiasm. A few of his favorites include: the almond tree, known for its delectable nuts; the gum tree, that is used to treat stone bruises caused when children often run around in their bare feet; black sage Roy says is almost impossible to transplant; Jamaican enemy that sports a pretty yellow flower; whistling pine wood that was used for the flag poles; and the snake tree which grew by cottage #9 and formed a spectacular arch over the road. A favorite of many residents and guests are the plumeria trees, more commonly known as the frangipani trees. The hotel owned by Sir James was named after this tree and they are inhabited by the spectacular frangipani caterpillars famous for their size and vibrant colors. Although all the trees gracing the island are beautiful in one way or another, either for their visual appeal or their usefulness, everyone ends up having their favorite, and Roy believes Haze's favorite tree was the coconut palm.

ISLAND VEGETATION AND TRANSPORTATION

Two Frangipani tree hosting colorful caterpillars and birds

Whistling pine along West End Beach

PSV also maintains a garden since it strives to offer the finest in vegetables and fruits in its restaurants. Roy, as well as many other gardeners over the years, has worked very hard growing vegetables in raised beds such as beets, lettuce, cherry tomatoes, spinach, and cucumbers, as well as various herbs including lemongrass, mint, and oregano. The island's fruit trees include lime and mango as well as the orange trees that have been planted in honor of Haze Richardson, Alice

Boss, Lee Blissett, and David Maxwell. Experts have routinely been brought to PSV over the years to help teach the staff better ways of growing these crops. This has helped the crew better understand soil quality and growing techniques to further enhance what was already being grown and offers suggestions for future harvests.

Raised beds of vegetables in the garden

The source of water on the island when men first arrived came from three ponds according to Goatie. The first pond, which was discussed in an earlier chapter, was used by Goatie and other men from 1966 to 1968 and its water was purified with the use of ash in barrels. Goatie believes the early settlers' water would have been from this pond. Water lilies have since been planted there in order to keep the mosquitoes at bay.

Lilies on water pond

Another one situated closer to the beach was probably spring-fed from an underground source but is no longer there. He thinks the inhabitants from Petite Martinique would have used this pond as their source for fresh water as it had a lot of sand on the bottom, which is considered to be a natural filter. It has since dried up. The last pond sits higher on the hill and is named Betsy's pond after a former coworker. It naturally fills with water and ditches were installed so the water would drain into the first pond. Today, modern desalination equipment is used to create fresh water that is collected and stored in large tanks.

Roy thinks it is important to be able to use rainwater to nourish the plants rather than fresh water from the desalination plant. There are plans in the future to store the rainwater for uses around the island rather than letting the water go to waste. These ideas, along with many others, are what make the staff special at Petit St. Vincent. The gardeners and landscapers are always working hard to make sure the island not only thrived in the past and continues to flourish now, but that the future will hold great possibilities as well as a home for guests and for themselves. Thanks and appreciation goes out to the landscaping staff for making PSV the breathtaking island it still is today.

ISLAND TRANSPORTATION: THE MOKES

"Can you call me a moke," old time guests will say; or new guests to the island are more likely to ask, "What the heck is a moke?" The moke is the official 'car' of PSV! It is sometimes referred to as a 'mini moke' which was originally constructed in England for the British Motor Corporation. [The word 'moke' is from ancient dialect meaning donkey.[9]] Sir Alec Issignois' design was a prototype for a Jeep-like vehicle but was originally engineered for lightweight military use. When that did not gain much appeal because of its relatively low clearance, it was offered to the general public in various countries as a utility car or beach buggy. It is amusing to note that when buying one new, the original options available were seats, heaters, and windshield wipers. If you did purchase these 'extras,' the proud new owner had to fasten them on themselves!

Cole Beadon's redesigned mokes

The first moke came as a whole unit from England and was basically used for room service. It was an all-steel construction which began rusting away. But before it fell apart Haze decided they should replicate it in fiberglass, using the original Moke as a mold. That method required it to be made in several pieces, which were later joined together and produced a somewhat rough finish. Unfortunately some of the critical joints would frequently come apart and have to be repaired, so Cole Beadon from England redesigned the moke and constructed a new mold. The main body could now be molded as a single piece, making it much stronger and more suited to fiberglass construction.

"Haze didn't want it to look too different from the ones in the brochures, so we kept the overall shape very similar. We made the molds and built the first new Mokes in one of the warehouses at the back of PSV – Haze shipped in the fiberglass materials from the US, and we bought and shipped in new and used parts and engines from the UK. My wife Beth and one of the PSV staff from St. Vincent named Tony, helped with the construction. The redesigned body included rounded edges instead of sharp ones, a six inch increase to the internal width, and a similar increase in overall length. Like the originals, the new Mokes were all light blue, with one notable exception – a mustard yellow 'convertible' built for Jennifer. However a couple of years later it was painted blue to match the others."[70]

Robert Hinds, better known as Blondie, has been resident mechanic since joining the PSV team in 1989. He has helped build more mokes and now maintains them as well. In total, PSV has had eight mini mokes with two originals still gracing the island. Blondie said that although the last one was built twenty years ago, he remembers how hard, and painful, it was. He worked with Reynolds to do the fabricating.

The molds first had to be coated with a light-blue gel coat [white was used for the roof, seats, and grill]. Then various layers of mat and woven fiberglass were added and wetted out with the resin until the required thickness was reached. The resin had to be mixed with just the right amount of a hardener and applied immediately so that it didn't set before they were finished. Blondie remembers the fine fiberglass splinters generated from sanding and cutting; these invisible pieces would get onto his skin and hands and were very painful. Although he wasn't working on PSV when the first moke arrived, he figured out that at least one of them had originally been yellow in color. He noticed this while sanding one after it had 'a little accident.' [That story a little later!] They all currently carry either an 850cc or 1000cc Mini engine with a four-speed standard transmission. The mechanical parts which include the engine, transmission, sub-frames, suspension, wheels, brakes, and steering are all imported from the UK. The tires ride on 10" rims and usually last only about two months since the smooth concrete roads can be rough on them. Although small, Blondie says they can get up to 90mph . . . and how that is known shall remain a mystery! However, that leads to the next story: Even with the limited amount of transportation on the island, there have been two moke crashes reported!

The first one involved Mr. Doc Burtell, who, while driving his mini moke in the vicinity of the dock house, crashed into a coconut tree and smashed the moke to death. It is not known to anyone except for Doc whether the accident was caused by failing brakes . . . or perhaps excessive speed! But either way, although he was not injured, the moke was a total disaster. Luckily it was rescued by some loving PSV guests from England; Jane White decided to buy it for her partner, Stephen Bullock, for his 50th birthday. They arranged to purchase the moke as is, and "after some miraculous body work by Blondie in the PSV workshop," they had it shipped back to England where it was restored to looking brand new!

ISLAND VEGETATION AND TRANSPORTATION

"It was a major project involving Jane tracking down the crashed moke via Roland and Blondie and then . . . got the body of the moke shipped to the UK via a Geest banana boat. Jane then had to find a donor mini moke vehicle in the UK with an engine and all the running gear and then had a mechanic 'transplant' the donor engine and rebuild as a finished moke." [11]

Wrecked moke

Refurbished moke with new owners Stephen Bullock and Jane White

All this was done as a total surprise to Stephen! The owners are clearly very happy with it as the picture indicates! They refer to it warmly as *Mook* because the shipping crate it was packed in had the words 'Dr. Jane and Mook' scrawled on the outside of it! "We are massive PSV fans and have been there every year for the last twenty years and in fact will arrive as usual [this year to] stay in cottage #4 which is our second home." [12] They say it is a joy to see many of the staff when they return such as Jeff, Roland, Bequia [Ezekiel Bess], Goatie, and Mattie to name but a few. "We are delighted Phil seems to value PSV for the same reasons and has kept the magic intact and improved it further. We are grateful to him we can continue to enjoy the island and its natural beauty," Stephen and Jane pointed out.[13]

The second reported accident on the island with a mini moke involved Otnel who was driving the vehicle and he unfortunately did not see a fast approaching coconut tree! The resulting crash produced a topless moke which forever remained as such and is now the only 'convertible' on the island! It became Jennifer's favorite vehicle and liked it so much she wouldn't let the top be put back on! Needless to say, the coconut trees are very special in this part of the Caribbean . . . they seem to be magnets, attracting mokes without any reason for hating them so much! As Blondie discovered when he was repairing the damage caused by Otnel's slight inaccuracy in steering, this was apparently the 'yellow Moke' that had originally been built for Jennifer.

Jennifer Richardson's moke

The mokes have not only offered transportation to the guests and staff but have provided many hours of enjoyment as well. Although this was designed as a military-use vehicle, it certainly has contributed greatly to PSV and they have served their time well. The original question of "What the heck is a moke?" is now more likely to be replaced with the request, "Bring around a moke! Let's go for a ride!"

CHAPTER TWELVE

PETIT ST. VINCENT
THE GEM OF THE CARRIBBEAN

To adequately understand a place such as Petit St. Vincent, people might seek advice from magazines, books, and websites to view other's opinions of the resort. There are travel writers galore who scan the globe for the best 'new place', the 'hottest location', or the 'gems of the world' and excitedly report their discoveries. PSV has acquired a profusion of awards over the past forty-six years. Their most recent include the *Andrew Harper's Hideaway Report* ranking them as 'World's Best Top Twenty International Hideaways' for 2006 and 2007 and PSV joined the ranks of *Condé Nast Traveler's* Gold List for 2008, 2009, and 2014 as well as its Hot List for 2012. As of July 2013, Petit St. Vincent proudly flies the flag of the *Small Luxury Hotels of the World* and is a *Visa* Luxury Hotel Collection member. The list of awards goes on and on. But awards signify only part of the reason people chose to come to PSV.

Petit St. Vincent is known for its solitude and guests can have as much or as little of that as they desire. Many guests arrive and chose to spend their time on the island alone with a good book; many others want to fill their days to the brim with excitement. Whichever suits your fancy, PSV offers many chances for seclusion, relaxed activities, or downright thrilling adventures each day if you are so inclined to partake.

Jigsaw puzzle highlighting Petit St. Vincent

Relaxing on cottage's private hammock

Simple pleasures on the island are many. They include strolling leisurely around the island or taking either of two walking paths which climb to incredible views. The first hike ends at the top of Marni Hill and the other hike ascends to the peak of Telescope Hill, named as the lookout location for enemy ships. Keeping in shape is easy if you climb either one of those two every day! Both are beautiful places to enjoy the sunrise or sunset and are perfect places for photo opportunities. Wonderful memories of guests and staff alike have been created there. Another tranquil spot would be to take a short boat ride to Mopion Island where you can relax and have a delicious picnic lunch.

This way to Marni Hill

View from Telescope Hill of PSV harbor and Petite Martinique

View of the Atlantic Ocean from Telescope Hill

Two yoga pavilions are located on the island for exercise and meditation which makes for a very nice start to the guests' day. Or there is a pavilion stretching out to the sea for relaxing and reading a good book. Bird watching and photography also rank high on the list of enjoyable leisure activities to do. In the Caribbean, there are approximately five hundred sixty different bird species and one hundred forty eight of them are endemic to the islands.[1] With so many different species to see, it is a bird watcher's paradise; and for the photographer, thousands of images are just waiting to be created.

Pavilion jutting out to the sea

For those guests with a little more spunk, jogging is a fine pastime. Either on the paths or on the sand, there are a couple of miles of winding routes to enjoy. Also, the tennis court is a great place to play a nice game and enjoy each other's company. Many a men have kept peace in their marriage when their wives 'magically' beat them at a match! Or at least that's how the story goes. Maybe it's the other way around! The staff members sometimes use the tennis courts to spend some free time. Quite the rivalries exist between them when they are in the heat of the game.

The tennis court was built in the early 1970s. It had to be resurfaced a total of five times over the years because of underground water damage from Telescope Hill runoff. It was finally decided in 2002 that the whole court must be redone with drains set under, and surrounding, the immediate area. These ditches were lined with stone and covered to look attractive. Then a six inch concrete slab was poured on top of already poured concrete beams. It was covered with an *Omni Court* synthetic grass surface. Incredibly, all of this was completed in a two month period while the island was closed for the off-season. Fortunately, materials were purchased rather than having to create them by hand as in the original construction days. Nineteen tons of sand was brought in, boats were chartered to bring in aggregate from St. Vincent, and a cement mixer was rented. Although quite a project in itself, having modern tools was definitely a plus this time around!

Of course, for those that love the water, swimming is a great way to spend the day. You can either let the current carry you along or get a great workout by swimming against it! And the dock house is full of fun toys to keep most anyone entertained. There are noodles to float away the hours, snorkeling gear for the more adventurous to see the wonders below the surface, kayaks and paddleboards for those with balance, and Hobie cats and windsurfing boards for the most adventurous. All of these are offered to the guests for their enjoyment along with instruction if needed. If for some reason the lessons do not make an expert out of you at once, the dock house staff always keeps a close eye, watching where you might venture off to. They have many stories about the rescues they have had to make over the years!

Windsurfing Fun

Kayaking anyone?

Paddleboarding the day away

Taking a Hobie Cat around the island

Snorkeling for treasures

For those who enjoy a leisurely outing, sailing opportunities are available to the Tobago Cays or other neighboring islands where guests can snorkel and swim with the sea turtles, or for those who like a bit more speed, power boats can be taken out on deep sea fishing expeditions. There is something for everyone on the island and Mr. Nichols, as well as the new owner Phil Stephenson, has guaranteed the fun will always continue . . . for those who want it!

At the end of the day, a private dinner is awaiting any guest who would like to have a romantic and quiet evening away from the main restaurant or beach bar. These private dinners have changed from prepared meals originally being offered, to a spectacular meal now being prepared by a private chef tableside at a location of their choice such as the Windward dock, the yoga pavilion, or on the west end beach. How romantic!

Franky Syukur making sure table is ready

Romantic sunset dinner on PSV beach

Excellent service awaits guests

After a splendid dinner, it is back to the cottage for some well-deserved rest and relaxation. Whether the choice was to lie around all day allowing stress to melt away or to have a day full of adventure, guests know that a beautiful cottage awaits their tired bodies and the sounds of crashing waves will be what lulls them into a peaceful sleep.

PETIT ST. VINCENT REGATTA

In years past, many have chosen to enjoy PSV by participating in the PSV Regatta and Fishing Tournament which was a highlight of the off-season. It was started in 1970 as a way to bring business to the island and attracted people from all over, with many competitors coming from Trinidad and Barbados. Originally this four day event was held over the Thanksgiving weekend, beginning on a Thursday and ending on a Sunday. It was a time for all to relax and enjoy.

PSV Regatta Brochure 1974

LETTER OF WELCOME

On this our fifth annual Petit St. Vincent Regatta and Fishing Tournament, we would like to warmly welcome all of our friends in the fraternity of men who share a keen interest in the sea. Each year it has been increasingly satisfying to observe the intense competition and fine sportsmanship which this Thanksgiving weekend has fostered. We sincerely hope to see another great Regatta and Fishing Tournament again this year and hope that all of our past participants will be with us again.

We wish you good sailing, good fishing, good luck and a merry weekend.

Mr. & Mrs. H. W. Nichols, Jr.
Owners of Petit St. Vincent

Letter of Welcome to participants

Prior to the start of the regatta, a booklet was available to would-be participants. In 1974, it included a welcome letter from Mr. and Mrs. Nichols, a yacht entry form, an anglers' entry form, an accommodation reservation form and some information about the four previous year's races. The yacht entry form was to be filled out and sent to the Petit St. Vincent Race Committee in either, St. Vincent, West Indies or St. George's, Grenada, West Indies. Which location you mailed it to would depend on the yacht's country of origin. Note that the addresses were listed as West Indies as this was before the country of St. Vincent gained its independence from Britain. The entrance fee was EC$15.00.

The accommodation reservation form in 1974 included the following information:

Petit St. Vincent will allocate a limited number of cottages for the housing of participants of the Petit St. Vincent Regatta and Fishing Tournament. Accommodations will be in short supply due to the large number of people expected to attend, and we therefore ask that interested participants fill out this form and return it as soon as possible. The charge per person per night for a bed only will be $10.00 E.C. or T.T. Food and drink will be extra. This price does not guarantee privacy (i.e. guests will share cottages). However, groups of six or more wishing to reside together please indicate this on the form and they will be assured a cottage to themselves. This arrangement applies only for the dates of November 28, 29, and 30, 1974. Petit St. Vincent will pick up any participants and their wives at Palm Island free of charge if they are arriving by air.

To enter the fishing tournament, the cost was EC$25.00 per vessel. Also included in the brochure's pages were fishing results from previous years so people could be reminded of their victories [or their losses!] Prizes could be won for the 'Biggest Fish Caught', the 'Highest Points per day per Angler', and the 'Highest Points per Boat'. In 1972, the biggest fish caught was a 55 lb sailfish on Friday and a 100 lb yellow fin tuna on Saturday. In 1973, the results were a 46 ¼ lb sailfish weighing in as the biggest fish caught one day, while the next day the biggest fish was a wahoo at 11 ¾ lbs! Besides fishing there were other contests including an underwater ecological treasure hunt and Sunfish sailing races. And for the land

ISLAND TREASURE

lovers there was the option of competing in tennis tournaments, volleyball matches, shuffleboard tournaments, or an on-island treasure hunt.

Within the brochure, a section called 'additional notes for all participants' highlighted the following points:

> *Toilets:* *This year due to the success of the beer drinking competition, the Management of P.S.V. has decided to install toilets in the vicinity of the Beach Bar.*
>
> *Beach Bar:* *Once again, serviced by lovely maidens, the Beach Bar will be in full swing – almost all booze (try our rum punch), cigarettes, matches, beer, etc. (excluding the maidens) will be available.*

In order to purchase food, drinks, ice, items from the boutique, and extra boat plaques, vouchers had to be used. They were actually called 'clams' and were for sale at the dock house. Doug Terman is credited for their creation and design. They were available in books of EC$25.00, EC$15.00, and EC$5.00. The vouchers could not be redeemed for cash but any clams a person didn't use could be put in donation boxes located at either the pavilion or the dock house and the equivalent value went to the *Save the Children's Fund* as a gift from *The Mariners of the West Indies*.

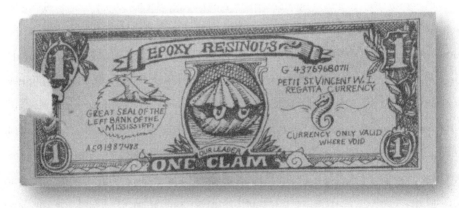

PSV Regatta Currency – the Clams!

The sailing instructions which were listed for Thursday were from 'St. Vincent to Petit St. Vincent' or from 'Grenada to Petit St. Vincent.' The race began at 08:00 hours and officially ended at 17:00 hours. On Friday the instructions were very specific and the course was set as:

Yachts must cross the Starting Line in a northerly direction and leave "Mopion" a sand cay with palm trees to the west of Petit St. Vincent, to Starboard thence proceed to the eastern coast of Prune Island, leaving Prune to Port, and thence to Union Island, leaving Union Island to Port and return to the Finish Line leaving "Mopion" to Port.

Mopion Island

The start began a bit later on Friday at 11:00 hours and ended at 16:30 hours. Prizes were handed out each evening at the PSV dock house. The last day of the race was the sail from Petit St. Vincent around Union Island. There was a caution noted in the sailing instructions for that day saying:

Caution should be exercised due to reefs on the east coast of Union, the west coast of Prune Island and Grand de Coi reef lying between Union Island and Prune Island.

The original barbeque for the regattas was in a pit down by the dock house area. There was always a suckling pig roasted or some other delicacy. A traditional steel drum band and a flute band entertained the crowd every evening! The regatta was a lot of fun for many people and stories are still being told about the excitement that was had during and after the races! Camaraderie at its finest! The last regatta was held in the early 1980s because the dollar of Trinidad became devalued and although their people were very big supporters of the race, they could not spend any money out of their country. It sadly ended a string of enjoyable years for many.

Other fun activities over the years that had been well-received included the weekly Wednesday night jump ups! The one held immediately before the regatta began each year was especially exciting as so many eager yachties were gathered, waiting to see who will be the big prize winners.

Entertainment over the years for the various jump ups, regattas, parties, and barbecues have consisted of musicians, singers, and steel drum bands. As previously mentioned, the first band was begun by the construction workers during the 1967-1968 season and was called the Island Waves. During the regatta they played and entertained all the racers. The Original Steel Band Orchestra is currently playing at the resort, wowing the guests with their exuberance and talent. They come as a group in a very small boat from Union Island to entertain at the beach barbecue which is held weekly at the new beach bar restaurant.

THE GEM OF THE CARIBBEAN

Island Waves Steel Drum Band arriving at the PSV beach bar dock

Member of the Island Waves Steel Drum Band setting up to entertain

Solitude, peace, tranquility, privacy, fun, excitement, and adventure are all reasons people initially choose PSV. The repeat guests find these e each time they 'come home', but highlights of seeing friends, many of whom are the staff, is what brings them back time and time again. It is not so much what Petit St. Vincent has to offer; it is what it doesn't have that makes it so attractive. And all the awards are just a reminder of that very fact.

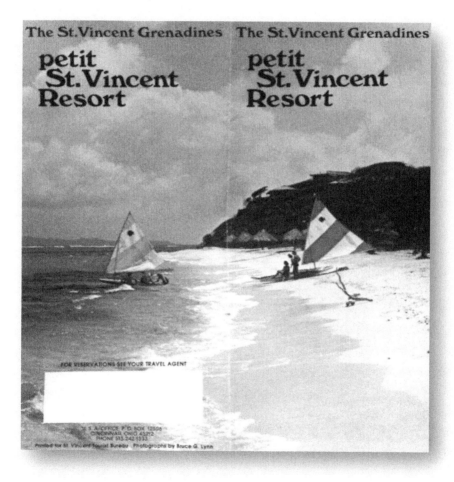

PSV early brochure 1974

CHAPTER THIRTEEN

THE BOATS OF PSV

Traveling to the Caribbean, one would expect to see sunny skies, beautiful white sand beaches, turquoise waters, and views that will take your breath away. And of course, Petit St. Vincent has all of these features. But one of the most important aspects of this island would have to be its boats! And PSV has had its share of them: sailboats, power boats, dinghys, and workboats, large and small, the island has had them all.

JACINTA

Jacinta is the name of the yacht, as you may recall, which Mr. and Mrs. Nichols chartered in 1964 and 1965 on their voyage to the Caribbean. It was a 77' schooner with three masts, was built in 1931, and designed by Samuel Crocker. Hazen Richardson and Doug Terman purchased her in New York.

Skippers of Jacinta . . . *Tony 'Stoney' Powell and Dave Corrigan*

Jacinta *under sail*

Jacinta *drifting down to Grenada with no wind and a broken engine*

Taking out old Buda engine in Trinidad

According to Tony Powell, who was the skipper of *Jacinta* during the 1969 season, the engine which is pictured "is the old Buda engine that died just before a charter, despite being prayed to every morning, so we had to do the whole charter without an engine!" Until it was replaced, Tony confessed they used to start the generator when they were sailing into a difficult anchorage so the guests on board thought the engine was actually running!! It was replaced with a "lovely Ford six cylinder which even had engine controls by the helm instead of a light switch by the helm."

Jacinta *Brochure*

ISLAND TREASURE

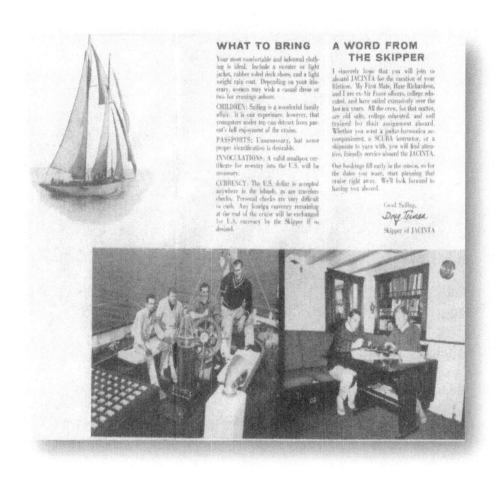

Original Jacinta *Brochure*

STRIKER

Striker was written about in a previous chapter, and as you might recall, is a 34' long sport fishing boat made by the Stryker Boat Company in the United States carrying two 108hp Ford diesel engines. Jennifer refers to her as 'our aluminum workhorse.' She was the first boat that Mr. Nichols bought for the island in 1967 and is still called 'Old Reliable.' "There were no comparable boats of this kind in the islands, as only sailboats abounded."[1] She would take guests on day charters to the Tobago Cays or Mayreau or on fishing expeditions. She could do it all! As you might recall, Haze and a Vincentian named Joe brought her down from Miami. They named her *Striker* after the boat builder's name.

Striker *at PSV dock*

Chester Belmar was her captain from 1968 to 1980 and she was used so often that she became rather well known. Haze personally would come to the rescue with *Striker* and pull many boats off reefs that had become stuck . . . mostly at night, of course. She was quite a celebrity back then, and since Haze did not charge a penny for his time and work, it helped quite a bit to promote her to 'superstar' status. After purchasing other boats over the years, *Striker* became more of a cargo boat than a passenger boat.

In 1979, an uprising occurred on Union Island. According to the *Bajan Reporter*, people of Union Island were suffering neglect by the St. Vincent government[2] and wanted their independence. Apparently, Revolutionary Prime Minister Maurice Bishop was to supply fifty Grenadian soldiers to support this movement for independence. Jennifer recalls during this time a boat was stationed off Palm Island waiting to attack. These men were dressed in police uniforms and because they thought PSV could be heavily armed, had supposedly decided to leave Petit St. Vincent to be dealt with last. "Little did they know we [only] had a flare gun," Jennifer recalls.[3] Haze soon found out that John Caldwell was holding them off with a 22 caliber rifle from the beach and when they retreated, not wanting their gas tank hit, Haze decided to try to overtake them. He boarded *Striker* and took off for Union Island which was the direction they were headed. He knew that since she had an aluminum hull, he would be able to run them down if necessary.

Jennifer stayed on shore with a portable VHF radio to serve as a lookout and a means of communication. She decided she better give any directions in French, just in case anyone was listening. However, her command of the language was not that fabulous so when she said, 'Regardez á votre droit and regardez á votre gauche,' she really didn't know which was left or right, so she would switch to English in mid-sentence. By that time, Haze was so confused that it wasn't much help! But being the easy-going man-under-pressure he was known to be, he calmly radioed to her, 'Ok Agent 007, just tell me in English . . . it will be easier on both of us,'[4] Jennifer recalls. It was a pretty scary situation but Haze keeping his cool was priceless to everyone involved.

Although he never had the opportunity to catch up with the invaders, *Striker* was invaluable that day and a hero in many people's eyes. They could count on that boat to do any job, even military-type work! She continues to be a blessing to the island and is still used today to transport workers to and from Petite Martinique.

WAKIVA

Wakiva was the very first boat Mr. Nichols bought to transport guests and was the second for the island. It was a 42' Grand Banks boat that was purchased from Mr. and Mrs. Beaufrond. They were the original creators of the Anchorage Yacht Club, the airstrip and the dock built on Union Island. Kate and Willis named the boat *Wakiva* because Kate's grandfather, Lamon V. Harkness had owned a steam yacht in 1907, built in Scotland for family cruising, and was named *Wakiva II*.

Wakiva *at the PSV boat dock*

Haze decided in September 2000 improvements were needed for *Wakiva*, so he shipped her off to Trinidad to make the necessary enhancements. Each job led to another and after more than a year's time, *Wakiva* was finally put back into the water. She had a new teak transom and deck, the cabin was refurbished, and new plumbing and electrical systems were installed.

However, in May 2003, a very serious problem surfaced. *Wakiva* was on fire! PSV boat captain, Maurice, who was at the Palm Beach restaurant in Petite Martinique at the time saw the flames and rushed over to the island because, from his perspective, he thought either the land or the dock house was on fire. When Maurice arrived, it was *Wakiva* that was going up in flames so he immediately called the staff kitchen from the dock house. Roland, the waiter, answered and immediately called Haze. Both of them came down to the dock with Jeff Stevens, boat captain of his own sloop *Jambalya*, not far behind. Jeff began pouring water onto the dock with the hose to keep the flames off of it. Haze wanted the water to be put on the boat instead but the small amount of water pressure was no match for the ever-increasing flames being fueled by the strong winds that night. When the bow line caught on fire, Haze told Maurice to unhook *Wakiva* from the dock so it wouldn't go up in flames as well.

As more and more people arrived, *Wakiva* slowly floated away into the harbor full of yachts. Jeff rushed back to his own boat to make sure everything was going to be alright with her but unfortunately she was headed directly at *Jambalya*. Jeff described *Wakiva* as "an inferno heading towards him."[5] She got pinned on the bow sprit [which to those non-sailors is the very front of the boat] and caught part of *Jambalya* on fire. *Wakiva* then drifted down the side of the boat and off into the harbor.

Jeff could feel the heat of the blaze on his face and decided there was nothing he could do; he jumped overboard to save himself. *Wakiva* was now completely in flames but incredibly it floated past every other boat in the very full harbor without causing the least bit of damage to any of them! No one actually followed her as they were afraid the diesel she was carrying could ignite, blowing up her and anything in the vicinity. Sadly, the fire completely destroyed *Wakiva* and she sank near Mopion Island. *Jambalya* had to have work done in Trinidad to restore what had burned but thankfully she was not a complete loss.

At first Haze believed an act of vandalism could have possibly caused the blaze, but it was later discovered the fire began from a fault in the new electrical system which had recently been installed. Maurice had just used *Wakiva* about four hours earlier that day and noticed some problems but they were minor and he was not alarmed that there was anything serious to worry about. Thank goodness the fire did not erupt when the guests he had picked up from Union Island were aboard.

HERA

Hera *at dockside*

In 2003 a brand new 43' Tiara built in Michigan was purchased to replace *Wakiva*. Haze bought her in Florida and decided to name her *Hera* after, of course, one of his dogs. She had much larger engines than *Wakiva*, with four times the horsepower. They were two Detroit diesel engines with 550 horsepower each. *Hera* was brought down from Providenciales in the country of Turks and Caicos in 2007. Loyal MacMillian writes of how Haze was such an 'amazing navigator' since he sailed with him to PSV on *Hera*. "He keeps a pad of paper with columns drawn down the pages. The pad contained information describing the route, distance, speed, and time to each destination. The pad was in depth, and very necessary for the success of our trip."[6]

ZEUS

Zeus was the next boat to grace the island. She was purchased in 1996 and it took twenty-four days to get her from Florida to the island.[7] *Zeus* was named after one of the labradors and was originally Haze and Lynn's personal boat. *Zeus* continues her work today, picking up guests from Union Island to bring them to PSV for their visit.

Zeus *with Captain Maurice displaying* Zeus' *official flag*

FREYA

Freya is a very famous yacht which is now gracing the harbor of Petit St. Vincent. As of 2013, new proprietor Phillip Stephenson is her proud owner. Phil uses this boat for pleasure sailing while staying on the island and she now has a very peaceful existence. However, her life was not always this tranquil.

She was built in 1963 by the Halverson family who originated from Norway. *Freya* was his most famous creation although he designed others. The reason she is so well known is the fact that she won the Sydney to Hobart Yacht Race three consecutive years in a row beginning in 1963! The race starts in Sydney, Australia where the yachts sail until they reach Hobart, Tasmania . . . a six hundred thirty nautical-mile trip. It is alleged by many to possibly be the most difficult race for yachts in the world.

John Corbett owned *Freya* for many years before Phil Stephenson purchased her. And as you may recall from a previous chapter, Freya had been caught in a storm one evening in the early 1980s. Cole Beadon, [the man who rebuilt the Mokes] recounts the incident as follows:

> *"I remember that although John had sailed through the pass between Punaise and Mopion hundreds of times, as he was approaching the cut under full sail, a rain squall blew through which reduced visibility. More importantly, the rain drops on the surface of the water broke it up, making it much less transparent so that it's almost impossible to see the colour changes on the bottom that normally warn you when you're running into shallow water and/or reefs."* [8]

The next morning *Striker* was brought out to pull her off the reef. *Freya* had been badly damaged and when she was pulled off, she sank, ending up in about twenty feet of water. Cole recalls "early the next morning John went out to see if there was anything they could do, and a local boat from Carriacou was already leaving the site, having just dived on it and stolen his sails – a very common occurrence in that part of the world where they consider any wreck fair game. John would later see his sails on one of the Carriacou boats."[9]

A week later Cole heard about *Freya's* sinking and contacted John. He had an idea how he could borrow a nine-hundred-gallon flexible water tank which belonged to the owner of a local saloon in Bequia to refloat her. With the help of two local lobster/conch divers [Squeal and his buddy], the operation of getting *Freya* back on the surface began. She had been underwater now for nine days.

> *"She was lying on her port side, so with the help of the two local divers the first thing I did was to put my spare deflated Avon dinghy in the forepeak and inflate it with a diving tank, which created enough buoyancy to bring her upright on the bottom. Then we took the 900 gallon water tank/air bladder down and dragged it into the main saloon. Even the tiny bit of air in it made it quite difficult to drag down ... We then proceeded to fill it using spare diving tanks, with one of the local diver's running back and forth to PSV to refill the tanks on their compressor. It took 12 tanks and then suddenly she started to rise. I can't tell you how exhilarating it was when she surfaced!"*[10]

They towed her back to PSV and beached her, damaged side up. Cole then gutted *Freya* and patched her with plywood strips and silicone sealant on the outside, and cement applied on the inside. *Striker* pulled her back into the water and Cole towed her to Bequia with his boat. John Corbett was on board *Freya* during the trip and Cole said, "I can still see [him] sitting there in the cockpit, steering as we towed her!"[11] Once in Bequia, Freya was repaired by an English boat builder named James Langston and was launched once again.

Cole Beadon getting ready to dive

Cole, John, and Squeal handling the large air bladder

Freya *breaks surface*

Freya *now ready to tow to PSV*

Arrival at PSV (Haze, white shirt, looking on) Piragua *in background*

Cole inspecting damages to Freya

Cole Beadon, John Corbett, and Squeal pumping sea water out of Freya

Back in all her glory ready to sail – John Corbett proudly presenting

Freya *racing back to PSV*

It is interesting to note that she was at one time connected to Petit St. Vincent and now the island's new owner has her in the harbor. She is quite beautiful and made of oak frames and fir planks. The decking material is fiberglass on wood. She has been repainted and some minor changes made since her star wins in Sydney, however she is just as pretty and fast as ever. John went to Australia one year and visited the Cruising Club of Australia which has a *Freya Room*, dedicated specifically to her.

With the ownership of boats comes the responsibility of taking care of them and there have been a number of truly remarkable captains throughout the years at Petit St. Vincent whom have done just that. The island has profited from their talent, time, and wisdom. They have worked long days transporting guests, taking them on day fishing trips and other excursions to neighboring islands and the Tobago Cays for snorkeling and other fun adventures.

Chester Belmar put in many years of dedicated service between the years of 1968 and 1980. He has decided to maintain his privacy with regards to his time at PSV, however it must be known that Chester is loved by many guests and staff members alike and his skills, sacrifices, and friendships are truly missed. Petit St. Vincent wants to recognize him for all that he contributed to make the resort a welcomed and enjoyable place for all. Chester currently lives on the island of Petite Martinique.

Maurice Roche began his career at Petit St. Vincent in 1986 and is still working on the island today, more than twenty-eight years later. He is originally from the neighboring island of Mayreau. His cousin James, who also worked on the island for many years, suggested Maurice look into working at PSV. James is a favorite of many return guests as well and currently has a restaurant in Mayreau. Maurice started his employment at the dock house, transporting people to Sandy Island or Petite Martinique, and taking care of the windsurfing and other recreation equipment. There also used to be additional work for the dock house staff selling water and fuel to the visiting yachts until PM put in their own fueling station.

James Roche picture at PSV and at his restaurant in Mayreau 2014

For about ten years, Maurice worked in the carpentry department on the island then moved on to crew the boats. Before Chester left to operate his own boat *Jahar* for island fishing trips, Maurice was able to train with him and with Chester's new replacement, Michael, to operate the larger boats. When Michael eventually left to start his own business, Maurice took over as boat captain. He has had a love of water and boats ever since he was a young boy on Mayreau; he used to build model boats and sail them on the salt lake there. He remembers how much fun he had and his affection for sailing has only intensified over the years.

Maurice now oversees the workers at the dock house and takes guests to and from Union Island. Although he enjoys all the boats, his personal favorite is *Zeus* because he can view a larger area of the sea from its high deck. However, he added, "*Hera* is a good sea boat because she takes the waves good."[12] When the resort closes, Maurice takes the boats out of the water for hurricane season. They are docked in Trinidad if they have major work to be done; if not, the boats will be hauled out in Carriacou. He usually stays with at least one of the boats until it is time to come back to PSV before the resort reopens in the fall. Maurice has remained on the island all these years because he loves his work and enjoys being close to his family on Mayreau.

PSV has always offered day charters on either sailing vessels or power boats for the guest's enjoyment. Bob and Marjorie [Marge] Law were the first couple to charter day sails for the guests on Petit St. Vincent. They built their sailboat *Trio* in Costa Rica in 1962 and left there in 1969, sailing through the Panama Canal. After a number of difficulties, they finally landed at PSV in late 1970.

Before joining PSV, Bob had been operating a sea plane from the port in Costa Rica where they lived to spot tuna for San Diego fishermen who were operating purse seine fishing boats. These boats did not operate trawl nets which pull along the bottom of the water catching fish, but rather use purse seine which is a fishing net that draws into a shape similar to a bag when closed to catch fish.[13] He would fly over the water looking for dolphins. Wherever there were dolphins, tuna would be found. He would then radio the fisherman their location and they would head out. Tuna are extremely large fish and would be sold by the pound. Bob was paid 5% of the total take for the day's work.

Bob met Haze in 1970 and although Bob was about ten years older than Haze they found they enjoyed each other's company because they had so much in common. Their connection came through flying because Bob's original job was a flight instructor in the Air Corps during World War II. Bob and Marge struck a deal with Haze to run the day sailing charters for the resort. They began that same year with *Trio*, their 35' Piver trimaran. [A trimaran is a yacht which has three hulls compared to a catamaran which has two.] They worked for two years on the island and then went to Vancouver, British Columbia, Canada to build another boat when *Trio* became infected with termites.

When they returned to Petit St. Vincent around 1978, the Laws again took over the sailing charters with their new boat *Pistachio*. It was a beautiful 55' Harrisdesign trimaran, which Bob and Marge modified to their exact specifications. After a day sailing with the guests, they often wanted to continue spending time with them and thought it would be nice to join them for dinner. Bob spoke to Haze about the possibility of them eating dinner at no charge in exchange for visiting with guests and answering any questions many had about the island. Bob and Marge noticed it was a strain sometimes for Jennifer and Haze to work hard all day and then stay up talking for a long time with the guests every single evening. This plan worked out well for all and was continued until 1980, when Marge's daughter Casey and son-in-law John decided to follow in their footsteps and take over the day chartering business for the Laws.

Pistachio

John and Casey Anthony began working for PSV in 1980 with their 38' Cross trimaran named *Piragua* and were very successful and well loved. "Both John and Casey and Casey's parents, the Laws, were huge favorites and friends of all of the guests on the island. It wasn't just an 'elective' to go on the day sail with either of these couples . . . it was something to which guests looked forward to, and lifetime friendships were formed between our guests and John and Casey," Jennifer recalls.[14]

The Anthony's had a nice arrangement with Haze; they offered a service of taking guests to the Tobago Cays, which they called 'the milk run,' but did not have to pay a commission to PSV. It was the same arrangement as Bob and Marge had when they were chartering. John did this so many times in the eight years they were at PSV, he said, "I could sail through the Mopion channel with my eyes blindfolded."[15] One year they were so busy they sailed thirty-one days in a row; an all-time record! Usually, however they averaged about eighteen days per month sailing and the other days were spent on the boat's upkeep, with maybe a bit of time off for rest and relaxation.

Piragua *under sail*

Their days were recounted as if it were yesterday that they sailed. They would welcome the guests on board and requested each one sign a register so they could always refer back to their names when they returned the next year. They would then sail to their destination where John would load everyone into a dinghy and motor to a wonderful snorkeling spot. Casey would remain on board to make lunch: A basket of food had been provided by PSV which usually consisted of chicken legs, cold cuts, a dozen eggs, a half pound of New Zealand cheddar, a wheel of camembert, mustard, pickles, and a nice selection of fruit. Casey would prepare the meat, make some deviled eggs and toss a tasty salad to serve the hungry guests when they arrived back on board after their morning in the water.

One particular year, a fin whale that somehow found itself trapped between Mayreau and the Cays when it was coming through the area from Bequia, was a highlight to show their guests. The Anthony's would anchor the boat close enough so that they could put speakers near the water for the whale to hear the Judy Collins' song, *Farewell to Tarwathie*, and the guests could snorkel nearby. This was very appealing as everyone seemed to enjoy 'swimming with the whale.' In those days, whaling was allowed [and actually still is today, their catch limited to two whales per year] however when the Bequia fishermen arrived to harpoon the whale, they refused saying 'it was put there by hands.' In other words, they felt it was too easy pickings for them so it was considered a trap. The fortunate whale was saved that day by divine help and eventually found its way out at the southern end of the area where it could continue on its journey.

The Anthony's usually sailed to the Tobago Cays but often, when there were returning guests, they were asked to go to different places. Chatham Bay, located on the western side of Union Island, was often their second choice of sailing destinations. John and Casey were able to throw in a line to see if they could catch a fish or two. The guests really enjoyed fresh fish for lunch as a compliment to the already delicious meal Casey prepared.

Another location for repeat guests would be a nice sailing trip to Salt Whistle Bay in Mayreau after lunch. John would take the guests on a walk up the hill to see the Mayreau Catholic Church which, besides being beautiful, has outstanding views of the Tobago Cays. Sometimes Casey would sail around to the harbor on the other side of the island while John and the guests were on their walk, and they would all meet up for their return sail. Another very popular location was White Island on the south end of Carriacou, which was the Nichols' family's favorite sailing spot, so the Anthony's would often take them there and would circumnavigate the island as part of their trip.

They got to know the guests very well and established long-time friendships with many of them. At the end of the day, oftentimes John and Casey would join their guests for dinner, just as Bob and Marjorie had been doing. John would entertain the guests with stories and information about the island which he especially liked since he has a friendly personality and likes to ask and answer a lot of questions! He remembers how the guests enjoyed the waiters when they gave their personal touch to the dining experience by reading the evening menu aloud. Because of the dialect differences, one particular waiter would pronounce certain dinner choices a bit differently than most guests were used to hearing it. For instance, *stuffed pheasant under glass* unknowingly became *stuffed peasant under glass* and *sweet bread* sounded more like *sweet breasts!* It was very apparent to the guests what the item really was so no one actually ever corrected the waiter; just a smile and a good memory of a job well accomplished!

It was Casey, however, who stole the show after dinner when she sang and played her guitar at the bar. She would 'perform' three to four times per week, entertaining the guests with her wonderful voice. It went over very well as she could offer selections from her very large repertoire of fifty to one hundred well-known songs. The guests loved her, the resort benefited, and John thought it was also good for the chartering business.

During the years the Anthony's were chartering for PSV, they had many fond and fun memories. For instance, famous singer-songwriter Bob Dylan sailed to PSV on his boat *Water Pearl* with his girlfriend soon after John Lennon had been murdered, John recalls. He remembers the story Bob shared with him about how he originally wanted to name his boat *Resurrection* but that his friend, who built the boat, said no because he was not a religious man. So that name was not going to work and *Water Pearl* was chosen instead. The Anthony's were pleased to have dinner with Bob one evening after which he was able to enjoy Casey's musical talent. The next day more excitement followed when John taught Bob how to snorkel off Mopion Island. John and Casey also remember Bob taking the time to come over to their boat *Piragua* to admire it. After the tour, John recalls Bob asking his girlfriend "Would you rather have a boat like this or a Mercedes?"[16] John was surprised to hear a question such as that and wonders which one she ultimately chose.

Besides being involved with their chartering business, the Anthony's were able to spend time with Haze, enjoying his company. They were amazed at his creativity and intelligence and could not imagine a person would be able to figure out all the logistics necessary to run an island, how to provide electricity when no neighboring island had power, and how to construct every building, all without having prior training or knowledge in these areas. They knew Haze was very bright and had benefited greatly from his education. John described him as a shy man who was a bit more comfortable speaking with the locals of the surrounding islands than with the guests. "He mostly stuck with business and was never really that open," John said.[17] He loved playing poker with Haze, and he, the chief of police in St. Vincent, and Bob Law were all very good opponents.

Haze told John many stories about his life prior to arriving in the Grenadines. Haze had been a navigator in the United States Air Force during the Cold War flying as a bomber pilot. He would fly over Western Europe with a sealed envelope in his pocket. It contained instructions on where the target was in Russia and where to drop the nuclear bombs he carried on board if ordered. Haze told John that no one knew the destinations that had been preplanned and that no envelope, thankfully, was ever opened.

John and Casey were very close to Jennifer as well and believe she deserves a lot of credit for making the resort look and operate the way it did. She marketed the island by writing to magazines and travel agents and would arrange 'fam trips' for them, as they were called, to familiarize themselves with PSV. In the Anthony's opinion, Jennifer was the one who got the island off the ground with her expertise and hard work. She added personal touches by keeping a black book with every guest's name and greeted each with a drink on the dock as they arrived. She would make sure there was a bottle of wine and fresh flowers in their room waiting to welcome them. She was responsible for decorating all the rooms, hosting the weekly evening cocktail parties at their home, and making each guest feel like the island was their own personal residence.

In their opinion, the boutiques were made successful by Jennifer as well because she was very talented in this area. She actually owned and operated others herself on the neighboring islands of Union and Bequia. Jennifer also helped John and Casey by promoting their boat. Although they worked very hard every day for their chartering business, they give much credit and respect to the Richardson's who they believed outworked everyone.

John remembers the staff very well and speaks highly of many of them. He thought Chester, the boat captain, "was very conscientious about guests and was one reason for the success of the place."[18] Chester's brother, Reynolds, who Haze called 'Pepper', was in charge of maintenance and is remembered by John as starting when he was just a young kid. And Slick was a great chef, Casey recalls, and he and John spent many hours playing ping pong. Depending on whether you ask John or Slick, you will find varying degrees of recollection as to who the best player actually was!

In the late 1980s when the island was available for Haze to purchase, John could tell he was under a lot of stress. He had to come up with the money "at a giveaway price," John recollects,[19] before the time limit ran out. When one of the repeat guests of the island was sailing with the Anthony's at that time, John thought it might be a possibility they could assist the Richardson's. He talked to them about the predicament Haze and Jennifer were in and suggested they get together and talk. This family ended up being the ones who worked out a deal to help the Richardson's buy the island.

Tragedy struck the Anthony's family when Casey's mother, Marjorie, died in May 1987. It was an especially hard time for Casey as she had their one year old son to take care of and was due in just a few short months with their second child. Haze was kind enough to fly them to St. Vincent when they decided to obtain permission to have her buried at sea. Three officials of the government were required to be present and soon it was arranged for her to be laid to rest off the coast of Bequia. They were very thankful that Haze and Jennifer were there to support both of them during their time of grief.

The Anthony's left the island in 1988 but remember their years there with fond memories of the Richardson's, the guests, and the entire staff. The friendships they made meant a lot to them and for many, the closeness still exists today. Many years later when the Anthony's two boys grew up, they visited the island and were given a tour by Diego, as they called Otnel. They had a great time on PSV and recalled how the north side of the island was filled with nurse sharks and conch. The boys understand now why their parents say their time at Petit St. Vincent was a most wonderful part of their lives.

Val and Marsha were the next couple who sailed the guests to their adventures in the Cays and various islands. Their boat was named *Camelot* and they stayed for a couple of years. Apparently, their boat had a piano aboard! After they moved on, Jeff Stevens took over offering the guests day sails. He is a long-time resident of the island, living in the adjacent harbor on his sailing yacht. Jeff came to Petit St. Vincent in 1991, with his boat *Friendship II*, when he heard the island was looking for someone to offer day trips. He went to make a phone call to Haze and Bob Law put him in touch with him . . . sort of. He remembers Bob saying, "Let me dial the number for you, Jeff" and after talking to Haze for about one hour, said, "Oh, by the way, hire Jeff Stevens for the day sails!"[20] Jeff credits his being a part of the island to Bob Law for this very, matter-of-fact reference.

Haze asked him to work while *Camelot* was away. He guaranteed Jeff work for six weeks in 1991 while Val and Marsha were to be away but he is still here today when they decided not to return. Jeff later decided to commission a new boat to be built, naming her *Jambalya*. She was a locally built boat handcrafted shipwrights on Carriacou. He went to Grenada to personally pick out the trees that would be used to build its hull. He describes how the shipwrights had to find the perfect shaped tree that would curve just the right way. After just a short time away, Jeff and *Jambalya* were back to PSV and have sailed guests to the Tobago Cays and neighboring islands ever since.

Jeff hails from the Isle of Wight, which is the largest island of England situated in the English Channel off the coast of Hampshire, where he lived until he moved to Turkey. His next move was to PSV. He always loved water and has made his home on a boat for over forty years. His current boat *Beauty*, a traditional Windward Island sloop, was locally built on the island of Petite Martinique. When asked why he continues to choose to have handcrafted vessels made rather than buying newer, more modernly constructed ones, he explained he wants to help keep the skills alive by supporting the local shipbuilders.

Beauty *under construction*

His trips include a beautiful sail to the Cays or your choice of destinations around the islands and a personal guide who takes you and your party to the beach or a fantastic snorkeling spot. While you are off the boat, he personally prepares a wonderful luncheon of steak, lobster or other fresh catch of the day, and chicken as well as a fresh salad! Quite a plentiful meal for those who worked up a hearty appetite swimming! He is known not only for his fine sailing but for his famous meals grilled right at the table for you. He was recently featured in *Forbes Magazine* as one of the 'Best Travel Moments in 2013'[21] alongside other spectacular adventures around the world such as climbing the mountains in Chile, swimming with dophins in Peru and joining in the celebration of Spain's Holy Week. The writer described the trip with Jeff on *Beauty* as "stuff of sailing fantasy."[22]

Beauty *under sail*

Taking guests to the Cays is one of Jeff's favorite trips and the one most often chosen. For many years, he had a special friend to show them when they arrived. He would anchor his boat in a particular place at the Cays away from the crowds and would go snorkeling, always taking a loaf of bread or something to feed the fish. Eventually, there was one that would come up to him every day he went there . . . it was a barracuda! Jeff named him, what else but, Barry! Jeff eventually was able to swim alongside him and Barry would eat out of his hand. Not quite so thrilling for some of the guests as Jeff would often leave this 'surprise' a secret until the barracuda would arrive! But many repeat guests couldn't wait to go visit Barry who was around for many years. Sadly, Jeff found out one day when the fish didn't show up that a local fisherman had killed him. He was angry as all the men in the area knew that was his special 'pet' and he still remains saddened today when he thinks of all the great times they had that were cut short unnecessarily.

Because Jeff has been on the island for so many years, he has many stories to share. He remembers one guest, who was on his honeymoon, was swimming in the sea. When he came out of the water he was devastated to find his brand new platinum wedding ring had fallen off his finger. He and many staff members searched a long time to find it but to no avail; the ring was lost forever. Jeff tried to console him by saying, "It's ok because you have left a little bit of both of you here and you can always come back and visit it anytime you want."[23] A lovely gesture for sure and it helped the man calm down just a bit. His worry now was what his new bride was going to say and if that warming thought would do enough to appease her!

Another interesting story Jeff tells is about how the yachts in the harbor were being vandalized after dark one year around 2005. Someone was going up to each boat on a small dinghy to see if anyone was onboard. The vandal would throw gasoline on deck and wait to see if there was any response when someone smelled the odor. If no one came to investigate, the person would climb aboard and steal things. Unfortunately, this went on for awhile and some of Jeff's friends actually had items stolen right off their yacht. One night Jeff saw the criminal and so he jumped into his dingy and chased him all the way to Carriacou. Unfortunately, he could not catch him as the thief had a larger boat with a bigger engine. The frustrating situation got so bad that the priest from the neighboring island put a curse on him. Apparently that worked because he was caught soon after and thrown into jail! And of course, wouldn't you know it; the boat the man had been using to commit these crimes was found to have been stolen. He had been coming over from Petite Martinique each night where he was living at the time to vandalize the yachts.

It appears that this was the only problem of its kind Petit St. Vincent has ever seen, which is a very good illustration of the island's distinctive qualities of peace and tranquility. In fact, there have never been keys to the cottages, just proving that in all these years, there are still no concerns. Sometimes guests or staff would find one of the yachtsmen wandering around and peeking into the cottages but, for the most part, they were just curious and had no bad intentions. Haze did have a confrontation one year with a Frenchman from a yacht, however. He was found by Haze to be looking in one of the cottages and was asked nicely to leave, explaining that the cottages are off limits to visitors since it is a private island. The man got angry at being told to leave and apparently told Haze he couldn't make him go. So Jeff remembers Haze replying 'wanna bet?' and went to call the police. Union Island officers soon arrived and took the man to jail for trespassing!

Jeff and Goatie recall another story, albeit very tragic, of a lady guest who died on the island. She was visiting PSV by herself in the late 1970s and had a rather nasty habit of drinking quite a bit of whiskey each week . . . to the tune of two cases in about seven days, Goatie recalled. She was found dead by one of the staff members with a case of whiskey sitting in the bathroom and another case sitting in the living room. Of course it was a very sad day for the island and a very scary one for the staff. But other than a couple unfortunate incidents in the forty-plus years of operation, the resort of Petit St. Vincent has never had a major problem.

Jeff sails guests to many locations around the islands from November through July. However, every year during the off-season, he arranges to haul his boat out of the water and put it up in Carriacou where there are hurricane moorings. And *Freya* will be joining *Beauty* as neighbors this year for the first time during the hurricane season. It can be quite a process since many things have to be taken off the boat in case of bad weather and a lot of preparation work needs to go into 'putting her to bed' Jeff claims. Hopefully nothing happens while he is away and can return to a boat perfectly intact. Many other sailors who do not wish to take their boats out of the water head for safer harbors such as those in Grenada, which is further away from the hurricane belt.

In November 1999, there was a hurricane that caused a problem for Jeff. Hurricane Lenny happened when he was building *Jambalya*. All the timbers to be used for its construction which had been brought up from Grenada were washed away from where they were being stored. Thankfully, he was able to find them and hauled them back. However, the twenty-foot swells that Lenny produced took roads away in Carriacou and damaged the beaches on PSV. The road down to the west end was completely wiped out. Retaining walls were built after that distressing day to hopefully avert problems in the future.

Jeff continues to enjoy his time at Petit St. Vincent. Many guests come back and request another sail with him, some repeatedly chartering his boat while they are on the island. Some want to go to the Cays or neighboring islands for a day's outing with lunch at a local restaurant, but many others just enjoy sailing with Jeff having no destination at all in mind. Travel writers who have sailed with him give

Jeff high marks for his friendliness, expertise, and wonderful meals he offers to everyone. The island is grateful for all the effort and time Jeff has dedicated to making PSV a great destination and another reason they feel they have just returned home.

Boats have always been an essential part of the island's existence. A great many stories, experiences, and fun times arise when people are around boats, own boats, or when they come in contact with a boat for the first time. Petit St. Vincent has had many a craft grace the harbor surrounding the island and all of them have provided people with not only a form of transportation but much enjoyment as well.

CHAPTER FOURTEEN

THE REMAKING OF PSV

The first time Phillip Stephenson heard of Petit St. Vincent was in 1986 when he was a young man of twenty-one. He was about to marry and he and his fiancée were trying to pick a honeymoon spot. Her best friend was also getting married and they were planning a honeymoon on PSV. When Phil's fiancée heard of this island, she also wanted to have it become their honeymoon destination. Phil asked how much it would cost because he was just a young law student. She said it was 'only US$400 per night' which to him at the time was a fortune. They ended up backpacking around Europe but the seed of PSV was planted in Phil's mind.

The second memory of the island was when he and his father chartered a sailboat in 1989. They regularly took week-long vacations together in the British Virgin Islands but one year decided to come to the Grenadines to try something different. They rented a 30' fiberglass sailboat in St. Vincent and while glancing at the *Doyle Cruising Guide*, noticed it listed PSV as a very nice but private island. When they reached PSV, they went ashore and were basically told to leave. He could not imagine then that he would someday become the owner of this island paradise.

Phillip Stephenson, otherwise known as Phil, was born in Texas in 1965. He graduated from the Kinkaid School in Houston in 1983 where he was governing council president and Texas state debate champion. He received his Bachelor of Arts in government graduating magna cum laude from Harvard College in 1986 and his Juris Doctor, again graduating magna cum laude, from Harvard Law School in 1990. Mr. Stephenson went on to become an adjunct professor of international finance at Georgetown University.

Between 1990 and 1991, he practiced corporate and securities law in the Washington, DC office of Baker & Botts working particularly on matters involving the Resolution Trust Corporation. Following, Mr. Stephenson was a senior official in the Office of International Affairs at the US Treasury Department during the years 1991-1993 where he worked on issues relating to the dissolution of the former Soviet Union, US participation in multilateral financial institutions and trade/aid issues with developing nations. He was the controlling shareholder in and CEO of International Equity Partners, L.P. from 1993 until its sale in 2001. IEP was a specialist manager of private equity and distressed debt funds in Asia and Eastern Europe from 1993 until its sale in 2001. And from 2002 until its sale in 2007, Mr. Stephenson was a shareholder in, and the vice chairman of, the Rompetrol Group, NV, which is an oil refining, trading, and marketing company operating across Europe.

Phil's current endeavors include: operates as chairman of the Freedom Group, a set of companies focused on investments, real estate and hotel development, and media; and serves on the Board of International Counselors for the Center for

Strategic and International Studies. He also is a member of the fellowship/community service organization PathNorth based in Washington, DC.

If there is any time left over after all of his business ventures, Phil works hard advancing the Stephenson Foundation which was founded to promote understanding, exploration, protection, and enjoyment of the world's oceans. Because of Phil's great passion of the sea, he became a certified scuba diver. Because his other interests include flying and sailing, he holds a private pilot license and owns three yachts, all of which are currently enrolled in the New York Yacht Club. The flagship is the 190 ton, 120' *Galileo*, a classic Sparkman & Stephens designed ketch built in 1989 by Palmer & Johnson, currently based in Palma de Majorca. He also speaks a number of languages including fluent French and conversational Romanian which come in handy while traveling the world.

Galileo *under sail*

In 2008, soon after Mr. Richardson's death, Phil was sailing his yacht *Galileo* for about two months from St. Martin to the Grenadines, making stops at all the islands. In April 2009, he came ashore to Petit St. Vincent. He walked up to the bar with his girlfriend, ordered a couple of rum punches, had a very nice time, and left for Grenada. He considered it a good experience. Then a couple months later, Robin Patterson sent Phil an email about an island for sale. Robin's presentation caught Phil's eye that day and he said to his friends, "I remember this place and I think I'm going to buy it!"[1] His friends shook their heads and thought it was just a crazy idea Phil had that would never happen.

Robin Patterson

Robin Patterson is a real estate developer with over thirty years experience in the international residential property market. He is a sailor as well of a 103' sloop, *Zanzibar*, and had sailed regularly in the Grenadines about three or four times each year. He came across PSV in 2006 and asked to meet with Haze. Robin explained to him that if he ever wanted to develop the island to please let him know. Robin had realized the chain of islands where Petit St. Vincent is located was mostly unheard of and many people were unaware of this 'hidden group' as he called it.[2] Robin wanted to put together the money to buy the island and asked if Haze wanted to go 50/50 on the purchase or to sell outright. Haze never was interested in pursuing this proposal.

When Haze's estate wanted to sell the island, Ms. Richardson's law firm of Baker & McKenzie, who was hired to represent the sale of the island as well as look for potential buyers, contacted Robin. [It is interesting to note that Phil used to be employed by that same firm!] Robin was referred to Phil by a contact in the UK, and ended up approaching him as well as nine other potential co-investors. Robin eventually heard back from many who were interested in the idea of purchasing the island but Phil responded in twenty-four hours. He understood the ideas Robin had for the island's development which were to renovate it entirely as he had done in other areas. Robin's inspiration for PSV was North Island in the Seychelles.

At that time, *Galileo* was in the Mediterranean moored at Calvi, on the island of Corsica. Robin, and a friend of Phil's named Colin Hart, flew to meet Phil and it was decided to purchase the island in the middle of a fifty-knot mistral wind. Phil was the majority investor providing all the cash required, and Robin and Colin were partners with sweat equity, advice, and work. The deal was closed that evening with a handshake.

Around the same time, Lynn had been approached by others who wanted to purchase the island as well. One particular buyer's offer was significantly higher than Mr. Stephenson's. She wanted to drop Phil's deal and go with the other, which was an unusually high bid in Robin's professional opinion. Lynn ended up switching contracts but after six months, the deal did not seem to be moving forward. Possibly realizing there was a potentially high risk of losing the transaction, Lynn met with Phil and the papers were finalized in Washington, DC in

November 2010. They bought the company that owns the island; all rights were transferred, including all assets and liabilities.

Prior to the closing, a due diligence trip was arranged in July 2009 between Robin and Charles Stephenson, Phil's brother who is an asset manager. They decided to advise Phil on what the island had to offer, the property's equity, what worked and what didn't, among other things. They also met with Lynn, who seemed very overwhelmed in the wake of Haze's passing. They reported their findings and in September 2009, while the island was closed for the season, Phil personally visited PSV. He remembers that he had a meal of rice and beans, it was very hot, and there was no air conditioning. Phil and Charles stayed in cottage #20 and clearly remember the details of what they saw: "There were two beds in the cottage, old furniture, heavy drapes, and wooden louvers which made the room very dark. The bathroom toilets had to be flushed at least three times to work properly. I thought the island was beautiful but the hotel was 'tired and in need of renovation.[3]

That evening, Phil had dinner with Lynn and remembers how she showed up in an elegant black dress while he was casually dressed in a pair of shorts. They had an excellent bottle of Puligny-Montrachet which Otnel later mentioned was also one of Haze's favorite wines. While there, Phil recalls meeting Goatie for the first time. Charles started to ask him detailed questions about the island and remembers Goatie just replying, "First you buy the island then I will answer your questions."

They took over the island on what Phil calls 'hurricane day' which was on November 1st when Hurricane Thomas hit the Grenadines. Colin and Robin came to the island to review the resort's needs; Colin's initial role was providing budgetary oversight and Robin served as informal general contractor on all projects. Eventually Colin Hart's responsibilities became more extensive. He is currently working with the purchasing, financial, logistic, and other general operations of the island and is on-site many weeks out of each year.

Meanwhile, Howard Penland, Phil's pilot, was busy as well, transporting people and supplies to the island in a 1977 Cessna 421 twin-prop plane. Phil's girlfriend Elena and their three-month-old son Jack soon moved to Barbados to live for a month while organization began on PSV. One of Phil's favorite stories was when *Galileo* came into Barbados, their belongings were packed up, and they sailed overnight for twelve hours to Petit St. Vincent. They dropped anchor and that is where Phil and his family lived during the time of the renovations. On one of his first evenings ashore, Phil walked into the restaurant to dine wearing shorts. A waiter came up to him and said, "I'm sorry sir but gentlemen are required to wear dress pants after 6:00 pm." Phil simply replied, "Not anymore" and proceeded to an available table.[5]

Letters were soon sent out to previous guests notifying them the resort was under new management and that renovations were already underway. Amenities that were going to be offered included bringing the island up to technological standards by offering television and internet services in all the cottages and adding a kid's club. Phil's original plan also included developing the west end of the island and selling small US$2 million homes and larger US$5 million dollar villas.

However, these proposed expansions to Petit St. Vincent did not go over well at all with previous and current guests. PSV was bombarded by letters in reply saying they wholeheartedly disagreed with these ideas and would not recommend the new ownership move forward in that direction as it would ruin the peace and tranquility they were trying so hard to find. Lynn Kendall, a long time guest mentioned in an earlier chapter, was on the island during this time and worked with Phil to explain how she viewed the proposed changes as well as how others felt as well. The plan for building more hotel rooms was cancelled as Phil quickly came to realize the unique nature and traditions of PSV.

Because of these modifications in the original renovation ideas, it was decided the island would stay open from November 2010 to May 2011. When it closed at the end of the season, two hundred workers were brought in as well as twelve foreign experts, such as architects, designers, plumbers, and electricians. An award-winning architect from Australia made one very important recommendation: He simply stated to Phil he wouldn't advise him to change a thing; the cottages are perfectly sited. This is certainly a tribute to Mr. Nichols, Haze, Doug, and Arne and therefore this is why the cottages remain basically unchanged today.

Robin's motivations for PSV came from his extensive travels, including the treetop spa idea from when he lived in Africa. Robin felt the resources of other islands would complement PSV as well. He had all the furniture and incidentals such as the pebbles for the showers, the tables, and crockery purchased in Indonesia. He especially loved the various woods; Dutch and Indonesian teak so rich in different colors. "PSV let me play with all these ideas," Robin said.[6] When they made renovations, they did things a bit differently from the original workers in 1966. They did not build complete stone walls but instead used stone facings, and for the electrical wiring, they snaked around the stone rather than putting wires within them as in the original walls. Robin says he is very passionate about Petit St. Vincent and because of this close feeling, he believes his interests and love of the island will always remain in his heart.

The projects ended up costing double the initial estimates but all new developments were implemented on time and the island reopened as planned in November 2011. This was intended to be a soft opening for friends and relatives to make sure everything worked and flowed easily . . . a chance to 'kick the tires' so to speak. The island's new modifications were a success; except for one small problem . . . they ran out of water!

Colin Hart says that no matter how much was accomplished during this twelve month period, it could not compare to what the Nichols, Haze and Doug accomplished. It was just a fraction of the incredible undertakings they began over forty years ago. Since then, many additional changes have been made. An exciting new system for making the island a bit more 'green' has arrived in the form of a new water bottling system. The water that is being made with a reverse osmosis system will be put into smart-looking glass bottles designed with the island's logo sandblasted onto the front of each. The guests will be able to enjoy drinking fresh bottled water from the island and will decrease or eliminate the need to purchase water in plastic containers.

New PSV water bottle

New renovations for the cottages

Goatie noted a few changes during the island's renovation. The solar still was taken out and the broken-up material was used as backfill to make the new beach bar as well as the adjoining boat dock. A new power station was installed, the freezer was expanded, and a storage room was added. New staff housing was built, but because of time constraints, the structures were not made out of natural island stone as are the existing buildings; but instead used a more conventional wood construction. The wall by the beach bar was made with a stone front rather than a full stone wall.

The beach bar was designed to make more room for guests to dine and to have a drink as well as to accommodate visiting yachties. Otnel and Goatie both highly approve of the bar because they believe it offers guests more privacy in the main restaurant and is spacious enough for many people, so there are no overcrowding issues. This is where the current Saturday night barbecues are held with entertainment from a steel drum band as well as weekly movie nights.

'Goatie's Bar' is the highlight of the area where yachties and guests alike can mix and mingle the night away.

Sunset from the beach bar

There was a new spa built which offers luxuries to the guests such as pedicures and manicures, as well as various massages including special deep tissue massages. There is also a meditation deck for those so inclined. For the visitors sailing into the harbor, the spa offers a welcomed indulgence.

There was a home built next to the owner's residence which is called the Bali House. It was originally designed to be a model for the homes that were to be offered for sale at Petit St. Vincent. It was constructed in Bali and then disassembled and shipped to PSV in containers. A gentleman accompanied the shipment to assist in its reconstruction although he did not stay until completion. Otnel noted this became a problem for the workers when trying to piece it back together. Today it is used as a guest house and is actually divided into two living quarters, each with a luxurious private bath.

Beauty *sailing past the Bali House*

The new owners have worked very hard making their acquisition of the island something to be proud of without compromising the aesthetic value of PSV from days gone by. Changes are not always a bad thing and what has been accomplished in these few short years is to their credit. Appreciation as well goes out to the hard working men who labored on this renovation, making what many said could not be accomplished in such a short time, a reality. Also great thanks are offered to all the guests who so caringly made their wishes, ideas, and suggestions known during the renovation stage. Phil Stephenson looks forward to a long and lasting relationship with PSV and wants to enhance the island for all to enjoy in the future.

Phil Stephenson with family and friends during 'Guys' Week' in November 2013

Phil Stephenson with Dr. Robert Ballard and Dr. Katy Croft Bell of the Ocean Exploration Trust. OET's research vessel Nautilus *visited PSV in both 2013 and 2014*

CHAPTER FIFTEEN

GREAT LOSSES

"The ache for home lives in all of us, the safe place where we can go as we are and not be questioned."[1] Maya Angelou spoke these words and they are befitting of the environment Petit St. Vincent has always offered. The island has been thankful for so many who dedicated part or even their whole lives to make this possible, working tirelessly for its benefit. They are forever in the heart and soul of the island and their efforts will never really die. However, a few of the people written about in this book have passed through this world but live eternally in our minds.

On February 17, 2001, Arne Hasselqvist, great architect of PSV, passed away. His death was not only sudden but tragic as well. He and his son were living in the Bahamas working on a project on Grand Bahama Island. According to the *Los Angeles Times*, a fire broke out in the duplex he was renting but he and Lukas were evacuated safely. However, they went back into the apartment to save the computers where all the designs and plans for the Bahama resort were stored. They never made it out alive.[2] What a tragic loss for the community of islands of such a great architect. Arne Hasselqvist was just sixty-three years old and his son Lukas only twenty-three years young.

John and Casey Anthony remember what happened prior to the fire breaking out in the duplex. "Someone had set Arne's car on fire. He went to put that fire out and when he returned to the duplex, Arne found that on fire as well," recalls John.[3] Prior to his death, Arne, Anita, and Lukas had been living on Mustique and were well known for the homes Arne created. His wife Anita ran the boutique at the famous Cotton House. He also had an international architectural design firm on St. Vincent. If it were not for him coming to the island and inquiring if Mr. Nichols needed an architect, the whole of PSV could have been completely different and his inspirations to the island would have never materialized.

He is deeply grieved for by so many who loved him and were grateful for his work. He is also missed by the people who never had an opportunity to meet him but have benefited from his creativity and genius in the architectural world. Petit St. Vincent was blessed to have him a part of its history.

Doug Terman, the man inspired to buy a boat and go sailing into the Caribbean, which started this whole adventure, lived a very full life. He was a pilot, then a sailor and manager of PSV, then later on in his life, a writer. He touched many lives while on the island and is remembered for the inspirations he had which brought fresh water to everyone working hard to build the resort. Without his ingenuity, Petit St. Vincent could not have had the tremendous start it did, nor would it have been able to continue being built into the distinguished resort it is today. Creativity and resourcefulness were Doug's gift to the island and PSV is forever grateful to have had such a person a part of its development. Douglas Terman sadly passed away in 1999 at the age of sixty-six.

A day in history where another great loss hit PSV was July 14, 1985; Harold Willis Nichols, Jr. passed away. If it was not for his love of the island, Petit St. Vincent as we know it would not have existed. He went through incredible trials and tribulations while waiting on an answer of whether or not he would be granted a license to own the island. But throughout those twenty months, he never gave up hope. Instead he made remarkable plans to put his ideas and desires to work and pressed forward creating the beautiful resort we all know and love. His and Kate's dream became a reality; Petit St. Vincent exists today because of his devotion to a small island in the Caribbean.

A little more than one year later, Katherine Nichols passed away as well. She was the backbone and strength Willis was able to draw from when times became frustrating. Kate raised four daughters, all while contributing her time and her love, not only to her husband and family, but to the resort as well. Her special touches enhanced PSV for all to enjoy even today.

Haze Richardson, as well, has passed into the next world, leaving us with only his memory. He and his wife were vacationing in Manuel Antonio National Park in Costa Rica in 2008. He was reported to have been swimming when a wave knocked him into a rock, hitting his head. No one was able to revive him. He is forever in the minds of all who knew and loved him. His importance in the role of manager and subsequently the owner of Petit St. Vincent cannot be denied.

Henry Ford once said, "Obstacles are those frightful things you see when you take your eyes off your goal."[4] Thank goodness Haze never stopped aiming for his target. He took PSV into his soul and created a home for himself and every guest who has ever been fortunate enough to have visited. He took time to see the good in the men and women who worked for him and he made a lasting impression which will be forever remembered in the hearts of so many. Hazen K. Richardson II, pilot, husband, PSV's administrator and proprietor, born October 1934; died February 2, 2008.

Each person, in a particular time and place in history, made a decision which led to them to their destiny. PSV is grateful for Mr. and Mrs. Willis Nichols, Mr. Hazen Richardson, Mr. Doug Terman, and Mr. Arne Hasselqvist who created our home away from home.

They will be forever missed.

THIS PSV

By Stephen Reed

Moody winds tease the crests of cyan waves,
Far beyond the island's rise.
Hurling snowy spray, bending sail and mast along their way.
Ragged energy that cannot break.
Yet as they cross the sandy shore, the island saps their mighty roar,
Turning gusts to meek-willed breeze,
Threatening gales to whispers that tease
Brown-dried leaves from shading trees.

Deep fathomed swells travel the surface of sapphire depths,
Far beyond the island's reach.
Those salt water steeps portend a power that never sleeps,
Rising in shallows to foam-topped strength.
Yet as they meet the hidden reef, the island steals their white-barred teeth,
Turning waves to rippled play
That glitter 'cross the turquoise bay,
The coral sand their final stay.

Weary travelers make their way by skies and seas from
Far beyond the island's peace.
They leave life's storm where turbulent air leaves souls atorn,
And time is lost to chaos' pace.
Yet as they light upon the dock, no longer life's a ticking clock,
From troubled hearts the spirit frees,
From tangled thoughts, the mind can see,
This calm oasis, this PSV.

SOURCES

Chapter One
1. Mock Up Brochure, Courtesy of the Nichols Family
2. Worms, L. (2004). The Maturing of British Commercial Cartography: William Faden (1749–1836) and the Map Trade. *Cartographic Journal, 41*(1), 5-11. doi:10.1179/000870404225019972
3. "Early People of St. Vincent and the Grenadines." *Global Literacy Project*. N.p., n.d. Web. 2 June 2014.
4. Lalueza-Fox, C.; Gilbert, M.T.P.; Martinez-Fuentes, A.J.; Calafell, F.; Bertranpetit, J. (June 2003). "Mitochondrial DNA from pre-Columbian Ciboneys from Cuba and the prehistoric colonization of the Caribbean". *American Journal of Physical Anthropology* (Wiley-Liss, Inc.) **121** (2): 97(12). doi:10.1002/ajpa.10236
5. Brizan, George I. *Grenada, Island of Conflict*. London: Macmillan, 1998. Print.
6. Logie, Phyllis. "The History of the Arawak People." *Humanities 360*. N.p., 25 Jan. 2010. Web. Summer 2014.
7. "Arawak History, Arawak History Museum, Carriacou." Personal interview. Feb. 2014.
8. Wallace, Wayne. "Role of Arawak Women in the Arawak Age." Centrelink, n.d. Web. Spring 2014
9. "Cuba Heritage .com - Cuban History, Architecture & Culture." *Cuba Heritage .com*. N.p., n.d. Web. 1 June 2014.
10. Tennesen, M. (2010). Uncovering the Arawaks. *Archaeology, 63*(5), 51-56.
11. "Early People of St. Vincent and the Grenadines." *Global Literacy Project*. N.p., n.d. Web. 2 June 2014.
12. Steele, Beverley A. *Grenada: A History of Its People*. Oxford: Macmillan Caribbean, 2003. Print.
13. "St. Vincent and the Grenadines." N.p., n.d. Web. Spring 2014.
14, 15. "Indigenous Peoples of the Americas" *Global Literacy Project*. N.p., n.d. Web. 2 June 2014.
16. Wallace, Wayne. "Role of Arawak Women in the Arawak Age." Centrelink, n.d. Web. Spring 2014
17, 18. Jones, David E (2007). *Poison Arrows: North American Indian Hunting and Warfare*. University of Texas Press. p. 29. ISBN 978-0-292-71428-1. Retrieved 2009-01-23.
19. Wiseman, Fred, Dr. *First Nations: Background* (n.d.): n. pag. Web. Summer 2014.
20. "Columbus, The Indians, and Human Progress." *Columbus, The Indians, and Human Progress*. N.p., n.d. Web. 21 June 2014.
21. Brizan, George I. *Grenada, Island of Conflict*. London: Macmillan, 1998. Print.
22. "HISTORY OF THE CARIBBEAN (WEST INDIES)." *HISTORY OF THE CARIBBEAN (WEST INDIES)*. N.p., n.d. Web. 2 June 2014.

23. "Columbus, The Indians, and Human Progress." *Columbus, The Indians, and Human Progress.* N.p., n.d. Web. 21 June 2014.
24. Logie, Phyllis. "The History of the Arawak People." *Humanities 360.* N.p., 25 Jan. 2010. Web. Summer 2014.
25. McD. BECKLES, H., & Shepherd, V. A. (1999). Chapter 9: Kalinago (Carib) Resistance to European Colonization of the Caribbean, *Caribbean Slavery in the Atlantic World* (pp. 117-126). JM: Ian Randle Publishers
26. Zinn, Howard, *A People's History of the United States*, Harper & Row, New York, 1980
27. McD. BECKLES, H., & Shepherd, V. A. (1999). Chapter 9: Kalinago (Carib) Resistance to European Colonization of the Caribbean, *Caribbean Slavery in the Atlantic World* (pp. 117-126). JM: Ian Randle Publishers
28. "HISTORY OF THE CARIBBEAN (WEST INDIES)." *HISTORY OF THE CARIBBEAN (WEST INDIES).* N.p., n.d. Web. 2 June 2014.
29. Claypole, William, John Robottom, and Coleridge Barnett. *Caribbean Story.* Harlow: Longman Caribbean, 1989. Print.
30. Steele, Beverley A. *Grenada: A History of Its People.* Oxford: Macmillan Caribbean, 2003. Print.
31. Letter to H.W. Nichols, from Nora Mantz, dated February 2, 1978
32. Letter from Lord Macaringy
33. Shepherd, Verene, Bridget Brereton, and Barbara Bailey. *Engendering History: Caribbean Women in Historical Perspective.* New York: St. Martin's, 1995. Print
34. "Slavery." N.p., n.d. Web, National Archives.gov. (Spring 2014).
35. "St. Vincent and the Grenadines." *Office of the Historian, St. Vincent and the Grenadines.* N.p., n.d. Web. (Spring 2014).

Chapter Two

1. "Petit St. Vincent History." Personal interview with Dwight Logan. (Spring 2014).
2. http://www.petitemartinique.com/church.htm by Dwight Logan
3. Roberts, Sherron. "Petit St. Vincent History." E-mail interview.
4. "Petit St. Vincent History." Personal interview with Doradine Ollivierre. (Spring 2014).
5. Vaccines. (2012, May 07). Retrieved August, 2014, from http://www.cdc.gov/vaccines/pubs/pinkbook/tetanus.html
6., 7. Roberts, Sherron. "Petit St. Vincent History." E-mail interview
8. Roberts, Sherron. "Petit St. Vincent History." E-mail interview and "Petit St. Vincent History." Personal interview with Doradine Ollivierre. Spring 2014.
9. , 10. The Barbados Museum & Historical Society, Today in History-Hurricane Janet. (n.d.). Retrieved April, 2014, from http://www.barbmuse.org.bb/2014/09/today-in-history-hurricane-janet/
11. Hurricane Retirement, Retrieved April 2014, from http://www.publicaffairs.noaa.gov/grounders/pdf/retirednames.pdf
12. The Barbados Museum & Historical Society, Today in History-Hurricane Janet. (n.d.). Retrieved April, 2014, from http://www.barbmuse.org.bb/2014/09/today-in-history-hurricane-janet/

Chapter Three
1-13. Nichols, H.W., journal entries, courtesy of the Nichols Family
14. PSV History [E-mail interview]with Elizabeth Marsh Nichols. (n.d.).
15-18. Nichols, H.W., journal entries, courtesy of the Nichols Family
19-21. PSV History [E-mail interview]with Elizabeth Marsh Nichols. (n.d.).

Chapter Four
1-2. Nichols, H.W., journal entries, courtesy of the Nichols Family
3-4. Dey, Richard. (209). Adventures in the Trade Winds. OffShore Press.
5. History of Palm Island, Retrieved June, 2014, from http://www.palmislandresortgrenadines.com/history-of-palm-island.htm
6-8. Nichols, H.W., journal entries, courtesy of the Nichols Family
9-11. (n.d.). *Paradise News, onepaper.com, July 26, 2000*
12. Dennis, Felix, *A Glass Half Full,* Retrieved June, 2014 from http://www.felixdennis.com/poetry/the-man-who-built-mustique-2/
13. History of Mustique, Retrieved February 2014, from http://www.mustique-island.com/about-mustique/history-of-mustique
14-18. Nichols, H.W., journal entries, courtesy of the Nichols Family

Chapter Five
1-2. H.W. Nichols journal entries, courtesy of the Nichols Family
3-4. Victory, N. (n.d.). History of the Island [Interview]. (March, 2014).
5. Nichols, H.W., journal entries, courtesy of the Nichols Family

Chapter Six
1-6. Johnson, S., Doug Terman Interview [Telephone interview]. (2014, October/November).
7. Powell, T. History of PSV Interview,[Email] (Fall 2014).
8. Johnson, S. Doug Terman Interview [Telephone interview]. (2014, October/November).
9-10. Richardson's Dairy History, Retrieved June 2014,http://www.richardsonsicecream.com.php53-7.dfw1-1.websitetestlink.com/our-history/
11. Daniel, Anne Richardson. History of Haze Richardson [Telephone interview]. (Summer 2014)
12-13. Richardson, J. [Telephone and e-mail Interview] (Summer 2014).
14. PSV Newsletter, December 1999
15-20. Richardson, J. [Telephone and e-mail Interview] (Summer 2014).

Chapter Seven
1. Victory, N. (n.d.). History of the Island [Interview]. (March, 2014).
2-4. Samuel, O. (n.d.). History of the Island [Interview]. (March, 2014).
5-7. Richardson, J. [Telephone and e-mail Interview] (Summer 2014).
8. Samuel, O. (n.d.). History of the Island [Interview]. (March, 2014).

9. Roosevelt, Theodore

Chapter Eight
1-16. Richardson, J. [Telephone and e-mail Interview] Summer 2014
17-22. Bethel, M. [Interview] Spring 2014
23. Cuffy, D. [Interview] Spring 2014

Chapter Nine
1. Morris, K. Chief Fisheries Officer report on PSV, July, 1987
2. Mitchell, J., Sir. (2014, Spring). History of Petit St. Vincent [Personal interview].
3-17. Corbett, J. [Interview] Fall 2014
18. Cole, R. (n.d.). Operation Urgent Fury. Retrieved August, 2014, from http://dtic.mil/dtic/
19-20. Corbett, John [Interview] Fall 2014
21. Beadon, C. [Interview] Fall, 2014
22. Tanase, Marin [Interview] Spring, 2014
23-30 Gaymes, Alfred [Telephone Interview] Summer 2014
31. Richardson, J. [Telephone and e-mail Interview] Summer 2014
32-33. Pringle, W. History of PSV [E-mail interview]. (2014, Fall).
34. Gaymes, Alfred. History of PSV [Telephone interview]. (2014, Spring).
35-41. Regis, R. (2014, Spring). History of PSV [Personal interview].

Chapter Ten
1. Nichols, H.W., journal entries, courtesy of the Nichols Family
2-7. Blissett, M. Petit St. Vincent Memories [E-mail interview]. (2014, Autumn).
8-9. Unnamed Guest. Memories of PSV [Personal Interview]. (2014, Spring).
10-16. Maxwell, A. PSV Memories, [Personal Interview]. (2014, Spring).
17-19 Kendall, L. PSV Memories, [Personal Interview]. (2014, Spring).
20-23 Reed, S. PSV Memories, [Personal and Email Interview]. (2014, Spring, Fall).

Chapter Eleven
1. Nichols, H.W., journal entries, courtesy of the Nichols Family
2. Leroy, L. [E-mail Interview], (Fall, 2014).
3. Sea Grape, Retrieved June, 2014, from http://www.gardenguides.com/74851-information-sea-grape-plants.html#ixzz34uk5ecrW
4. Mahogany, Retrieved September, 2014, from http://www.plantguide.org/mahogany-tree.html
5. Barclay, E. (2011, August 15). Coconut Water to the Rescue. Retrieved July, 2014, from http%3A%2F%2Fwww.npr.org%2Fblogs%2Fhealth%2F2011
6. Blue, L. M., Sailing, C., DeNapoles, C., Fondots, J., & Johnson, E. S. (2011). Manchineel Dermatitis in North American Students in the Caribbean. *Journal Of Travel Medicine*, *18*(6), 422-424. doi:10.1111/j.1708-8305.2011.00568. and Lauter, W. M., Lauretta E. Fox, and

William T. Ariail. "Investigation of the Toxic Principles of Hippomane Mancinella, L. I. Historical Review." *Journal of the American Pharmaceutical Association* 41.4 (1952): 199-201.
7. . Jones, David E (2007). *Poison Arrows: North American Indian Hunting and Warfare.* University of Texas Press. p. 29. ISBN 978-0-292-71428-1. Retrieved 2009-01-23.
8. Doyle, R. Island Vegetation [Personal Interview]. (2014, Spring).
9. Wood, J. (2005). *Alec Issigonis: The Man Who Made the Mini.* Breedon Books Publishing. ISBN 1-85983-449-3.
10-12. Beadon, C. [Interview] (Fall, 2014).
13. Bullock, S., White, J. [E-mail Interview] (Fall, 2014).

Chapter Twelve

1. Caribbean Endemic Bird Species, Retrieved July, 2014 from http://www.caribbeanbirdingtrail.org/caribbean-birds/caribbean-endemic-bird-species

Chapter Thirteen

1. Richardson, H., PSV Newsletter, November 2007
2. Union Island – The Revolution in the South Grenadines that Never Was. (n.d.). Retrieved August, 2014, from http://www.bajanreporter.com/2012/10/union-island-the-revolution-in-the-south-grenadines-that-never-was/
3-4. Richardson, J. [Telephone and e-mail Interview] (Summer 2014).
5. Steven, J. [Personal Interview] (Spring, 2014).
6. Richardson, H., PSV Newsletter, November 2003
7. Richardson, H., PSV Newsletter, November 2007
8-11. Beadon, C. [Interview] (Fall, 2014).
12. Roche, M. [Personal Interview] Spring, 2014
13. Call to Hold Talks over Trawl Net Ban. (n.d.). Retrieved Fall, 2014, from http://www.thestar.com.my/News/Community/2014/11/11/Call-to-hold-talks-over-trawl-net-ban-Boat-operators-to-face-millions-in-losses-says-MCA-Youth-chief/
14. Richardson, J. [Telephone and e-mail Interview] (Summer 2014).
15-19. Anthony, J. [E-mail Interview] (Fall 2014).
20. Steven, J. [Personal Interview] (Spring, 2014).
21. Abel, A. (n.d.). Adventures Of A Lifetime: My 12 Best Travel Moments Of 2013. Retrieved Summer, 2014, from http://www.forbes.com/sites/annabel/2013/12/02/adventures-of-a-lifetime-my-12-best-travel-moments-of-2013
22. PSV Newsletter, November 2003
23. Steven, J. [Personal Interview] (Spring, 2014).

Chapter Fourteen

1. Stephenson, P. [Interview] (Spring, 2014).
2. Paterson, R. [Telephone Interview] (Fall, 2014).
3. Stephenson, P. [Interview] (Spring, 2014).
4. Stephenson, P. of brother Charles Stephenson [Interview] (Spring, 2014).
5. Stephenson, P. [Interview] (Spring, 2014).

6. Paterson, R. [Telephone Interview] (Fall, 2014).

Chapter Fifteen
1. Angelou, M. quote
2. (2001, February 17). Arne Hasselqvist; Architect Created Celebrity Homes. Retrieved May, 2014, from http://articles.latimes.com/2001/feb/17/local/me-26577
3. Anthony, J. quote
4. Ford, H. Quote

PHOTO CREDITS

COURTESY OF:

COVER
Phil Stephenson

FORWARD
Sir James Mitchell, photo by Marcus Lyon

INTRODUCTION
Phil Stephenson

JENNIFER REMEMBERS
Jennifer Richardson

CHAPTER ONE
5 Google Images

CHAPTER TWO
22 Lana Shore mistyshoresphotography.com
23 www.wunderground.com/hurricane/atlantic/1955/Major-Hurricane-Janet

CHAPTER THREE
27, 29, 30 Hub Rail Magazine, 1979 winter
30 The Harness Racing Museum & Hall of Fame, Goshen, NY
33, 37 Nichols Family

CHAPTER FOUR
44, 48-49, 57-60 Nichols Family
50 Phil Stephenson
61, 62-64 Nichols Family

CHAPTER FIVE
Nichols Family

CHAPTER SIX
75 Seddon Johnson
76-77 Tony Powell
78-79 Richardson's Diary website: www.richardsonsonsicecream.com
80 Tony Powell
79, 85 John Corbett

CHAPTER SEVEN
92 Rosa Shore
92-94, 100 Lana Shore mistyshoresphotography.com

CHAPTER EIGHT
105 Lawrence P. Bemis
106 PSV
109 Lawrence P. Bemis
113 Jennifer Richardson
115 Malcolm Blissett
116 PSV
117 Lana Shore mistyshoresphotography.com

CHAPTER NINE
131 Jennifer Richardson
134 Beth Beadon
137, 140-141, 143 John Corbett
144 PSV
146-149 Beth Beadon
151 Tim Wright photoaction.com
152 John Corbett
168 Roland Regis
169 Lana Shore mistyshoresphotography.com
170 Phil Stephenson

CHAPTER TEN
175- 176 Malcolm Blissett
177, 179-180 Lana Shore mistyshoresphotography.com

CHAPTER ELEVEN
188, 190-191, 193-194 Lana Shore mistyshoresphotography.com
195 Lawrence P. Bemis
197, 199-202 Lana Shore mistyshoresphotography.com
204 Cole Beadon
207 Roland Regis
207 Bullock/White
209 Cole Beadon

CHAPTER TWELVE
213 PSV,
213-214 Lana Shore mistyshoresphotography.com
215 Beth Beadon
216, 218-222 Lana Shore mistyshoresphotography.com
223 PSV
224 Nichols Family
226 Otnel Samuel
227, 229-230 Lana Shore mistyshoresphotography.com

247 Malcolm Blissett

CHAPTER THIRTEEN
234-236 Tony Powell
237-239 Seddon Johnson
240, 242 John Corbett
244-245 Lana Shore mistyshoresphotography.com
248-252 Beth Beadon
253, 255 (right) Lana Shore mistyshoresphotography.com
255 (left) PSV
257-258 John and Casey Anthony
265 Malcolm Blissett
266 Lana Shore mistyshoresphotography.com

CHAPTER FOURTEEN
273, 275 Phil Stephenson
276 Robin Patterson
281-284 Lana Shore mistyshoresphotography.com
285 Phil Stephenson

AUTHOR PICTURE
310 Morgan Shore

INDEX

A

Anthropologists, *10*
Abercromby, Sir Ralph, *14*
American War of Independence, *14*
Anthony, John and Casey, *257-264*, *288*
Aquart, Joseph Modeste, *19-21*
Arawaks, *3-6, 8-11, 15, 190*
Archeologist, *3, 7*
Arrowroot, *9*
Artifacts, *5, 68*
Atlantic Ocean, *2, 31, 76, 181, 215*
Awards, *viii, 167, 212, 231*

B

Bajan Reporter, *241*
Ballard, Robert, *285*
Barbados, *vi, vii,, 11- 12, 20, 23, 55, 96, 107, 146, 162, 222, 279*
Barracuda "Barry", *267*
Baskets, *7, 158, 259*
Beach Bar, *vii, 93, 193, 226, 228-229, 282-283*
Beadon, Cole, *150, 204-205, 246-251*
Beauty, 264-266, 269, 284
Bell, Katy Croft, *285*
Belmar, Chester, *83, 108-109, 114, 142, 173-174, 205, 240, 254-255, 262*
Bennett, Thomas, *13, 18*
Bequia, *iii, 5, 11, 20, 98-99, 111-112, 127-128, 131, 135, 140, 150, 162-208, 247, 259, 262-263*
Bethel, (also see Bethel, Mattie), *18-19, 21-22, 33, 35-36, 89, 99, 107, 114-116, 142*
Bethel, Mattie, viii, 114-121, 124, 142, 159, 167, 173, 208
Blissett, Malcolm and LeClaire "Lee", *174-180, 192, 201*
Boss, Alice, *201*
Boutique, *vii, 104, 114-115, 226, 262, 288*
Bowie, David, *45*
Boyez, *6*
Brazil, *6, 118*
British National Archives, *15*
Brown, Joe, *42, 240*

Bullock, Steven, *221, 223*
Burtell, Doc, *221*

C

Caldwell, John "Johnny Coconut" and Mary, *43, 127, 192, 241*
Camelot, 264
Cannibal, *10*
Canoes, *3, 5-6, 8, 11*
Canouan, *18, 20, 120, 130-132, 162*
Carbet, *8*
Caribs, *5-10, 12-15, 190*
Caribbean, *i, iii, viii, 2-3, 6-7, 9, 12, 15, 32, 34, 45, 47, 70, 142, 150, 168, 180, 184, 189, 195, 208, 216, 234, 289*
Carriacou, *5, 18, 23, 31*
Chatoyer, Joseph, *14*
Chatham Bay, *259*
Ciboney, *3, 5*
Class Structure, *3*
Columbus, Christopher, *9-11*
Compton, John, *iv, 128, 131*
Conch, *7-8, 68, 106, 162, 247, 264,*
Corbett, John, *90, 133, 134-143, 145-150, 151-152, 246-247, 251-252*
Cotton, *4, 6-7, 9, 12-14, 20, 34, 45, 188, 289*
Cordice, Kenny, *153, 165*
Currency
    $Clams, 91, *226*,
    $EC, *69, 166, 225-226*
    $T.T., *225*
    $US, *v, 35, 43, 139, 165, 273, 279*
Cousteau, Jean-Michel, *iv*
Coyaba, *4*
Crops, *4, 7, 12-13, 201*
Crocker, Samuel, *75, 224*
Cuffy, Delia, *122-124*

D

Daisemon, 63
Danes, *2*
De la Rochette, L.S., *3*
Dennis, Felix, *46*
DaSilva, *89,*
Dey, Richard, *41*

DeGonville, Harry, *95*
Dominica, *12, 22*
Dowling, John Pius, Archbishop, *19*
Doyle, Roy, *189, 193,*
Duncan, *89, 93*
Dutch, 2, 11, *280*
Dylan, Bob, *261*

E

Eberhart, Bill, *46-47*
Emancipahon, *15*
Encantada, 76
Engendering History, *15*
European, *iii, 6, 9, 11-12, 14*

F

Faden, William, *3*
Fedeial Queen, 51-52, 93
Flag System, *54*
Fortunate Island, *8*
Forward Islands, *3*
Fishing, *6-8, 42, 52-53, 56, 162, 184, 220, 222, 225, 239-240, 254-256*
Frangipani, *iii, 5, 111, 127-128, 198-199*
Frederick, Daman "Dumbar", *65-66*
French, *2, 11-13, 168, 241, 268, 275,*
Freya, 134, 137-138, 152, 246-254, 269,
Friendship II, 264

G

Gairy, Eric, *iv, 128*
Galileo, 120, 275-277, 279
Garifuna, *11-12, 14*
Gaymes, Alfred "Slick", *156-164, 173, 263*
Genocide, *11*
German, Lord George, *13*
Geraldine Louise, 20
Goatie, "Noel Victory", *viii, 41, 49, 52, 63-64, 67-68, 83, 88-93, 95, 98, 101, 142, 167, 173-176, 183, 202, 208, 268, 278, 282-283*
God of the Sky, *5*
Goddess of the Earth, *5*
Grant, Mr., *143*

Grenada, *iii, 2, 11-13, 18, 23, 32, 34, 42, 44, 54, 65, 84, 97, 128, 145-146, 150, 184, 189, 224, 227, 236, 264, 269, 276,*
Grenadines, *viii, xiv, 2-3, 5, 13, 15, 27, 32, 34, 36, 47, 51, 105, 111-112, 127-128, 133-134, 162-163, 188, 262, 273, 276, 278,*
Guiana, *6*

H

Halverson, *246*
Hairoun, *11*
Hart, Colin, *vi, 277-278, 281*
Harvard, *11, 27, 274,*
Hasselqvist, Arne, *43, 45- 46, 127, 132, 288, 290*
Hawks' Bells, *9*
Hera, *116-118, 143, 244-245, 255*
Herbs, *200*
Hilfiger, Tommy, *45*
Hinds, Robert "Blondie", *205-208*
Hurricanes, *5, 23-24, 35, 45, 69, 98, 255, 269, 278*
Huts, *7-8, 178, 179, 193*

I

Igneri, *3*
Indigenous, *12, 192, 195*
Initiative, PSV, *iv, 12, 128*
Initiative, Willoughby, 1667, *12*
Island Activities, *52, 212-220*
Issignois, Sir Alec, *203*

J

Jacinta, *32, 40-41, 63, 65, 67, 69, 75, 77, 80-81, 234-239*
Jagger, Mick, *45*
Jahar, *144, 255*
Jambalya, *243-244, 264, 269*

K

Kendall, Jeremy and Lynn, *183-185, 279*

King, Elvis, *119*
King Ferdinand, *10*
King George III, *3*

L

Latifundia, *12*
Law, Bob and Majorie, *147, 149, 256-258, 260-261, 264*
Laws, *6, 15, 35, 40, 108, 157, 184, 273-274, 277*
Leeward Islands, *2-3, 12, 47*
Leroy, Linda, *189*
Lesser Antilles, *2-3, 9, 11-12, 15*
Lily Rose, *22*
Locals, *136, 164, 261*
Lord Macaringy, *13*

M

Maboya, *6*
Marni Hill, *2, 154, 214*
Mariners of the West Indies, *226*
Maxwell, Ann and David, *181-183, 201*
Mayreau, *18, 20, 98, 162, 240, 254-256, 259-260*
Messel, Oliver, *45-46*
Melvin, Howard, *47*
Mitchell, Sir James, *iii, 5, 40, 111, 113, 127-129, 131-133, 150-152, 161, 163, 198,*
Modyford, Sir Thomas, *12*
Moke, *53, 175, 180, 203-209, 246*
Morison, Samuel Eliot, *11*
Mopion, *109, 110, 118, 141, 214, 227, 244, 246, 258, 261*
Morris, Kerwyn, *129*
Morse, Robert, *13, 18*
Mustique, *iii, vii, 45-46, 96, 100, 127-128, 130, 132, 288*

N

Natives, *10*
Nichols, *13, 24, 26-37, 40-43, 45-48, 51-52, 54-56, 60, 64-65, 69-70, 74, 80, 84-85, 88, 101,* *127, 145, 150, 173, 188, 120, 124, 234, 239, 242, 260, 280-281, 289-290*
Niña, *10*
Nautilus, 285

O

Ocean Exploration Trust, *285*
Ollivierre, *20-21, 99*
Operation Urgent Fury, *145*

P

Palm Island, *31, 43, 192, 225, 241*
Paradise News, *45*
Patterson, Robin, *vi, 276-278, 280*
Penland, Howard, *279*
People's Revolutionary Government, *145, 150*
Petite Martinique, *18-23, 31, 33-35, 41-42, 51, 63-64, 66, 99, 106, 128-129, 135, 140, 145-146, 150, 155, 181-183, 202, 215, 241, 243- 244, 264, 268*
Pinta, *10*
Piragua, *6, 250, 257-258, 261*
Pistachio, *256-257*
Plantation, *12, 136, 140*
Poison Arrow Curtain, *11*
Polygamous, *6*
Pottery, *5, 68*
Poultice, *9, 198*
Powell, Tony, *234, 237*
Priest, *6, 18-20, 268*
Prime Minster, *iii, 5, 13, 111, 127, 132, 145, 163, 241*
Princess Margaret, *45*
Pringle, William "Billy Bones", *162*
Providence, *49, 89, 93*
PSVI, *176-177, 192*

Q

Queen Elizabeth, *111-113, 127, 153*
Queen Isabella, *10*

R

Reed, Penny and Steven, *183-185, 293*
Regatta, *91, 131, 134, 137, 145-146, 150-151, 222-226, 228*
Regis, Roland, *156, 164-168, 173, 207-208, 243*
Reynolds, *83, 114, 165, 205*
Richardson Dairy, *78*
Richardson, Haze, *iii, vi, vii, viii, ix, xiii, xiv, 32-33, 35-36, 40-44, 52, 54, 63-67, 73-75, 77-79, 80-85, 88, 90, 93, 96, 98-100, 104-111, 113-115, 117-118, 121-123, 128-138, 140-141, 145, 147, 149, 154, 157-167, 177-179, 181-182, 188-189, 198, 200, 205, 234, 240-245, 250, 256-258, 261-264, 268, 276-278, 280-281, 290*
Richardson, Jennifer, *xiii, 81, 83-85, 96, 98, 104, 106-115, 131, 157-158, 161, 164, 205, 208-209, 239, 241, 257-258, 262-263*
Richardson, Lynn, *114, 245, 277-278*
Richardson, Roland, *168-170, 196*
Ritualistic Beliefs, *4*
Roatán, *14*
Roberts, *20-21, 34, 190*
Roche, Maurice, *117, 241, 243-245, 254-255*
Rose, Dennis, *63, 67, 88-89*

S

Samuel, Otnel, *83-84, 90, 95-101, 141, 148, 156, 173, 184, 208, 263, 278, 282-283*
Santen, Harry, *40*
Second Carib War, *14*
Servants, *10*
Settlers, *12, 202*
Shell, *7-8, 68, 77, 106*
Shore, Lana, *ix, 310*
Solar Still, *36, 41, 64, 69, 282*
South America, *2-3, 5, 9*
Snorkling, *52, 185, 217, 220, 254, 259, 265, 267,*
Spa, *vii, 104, 180, 280, 283,*
Spain, *9-10, 12, 265,*
St. Hilaire, Mary Ann Constance, *18*
St. Lucia, *iii, 12, 42, 47, 128-129*
St. Vincent, *2-3, 5, 10-15, 18, 22, 35, 42, 46, 51-52, 54, 63-64, 66, 76, 95, 108, 111, 122-123,* *127-128, 133, 157-158, 161-163, 189, 205, 217, 224-225, 227, 241, 261, 263, 273, 289,*
Steel Drum Band, *51, 228-230, 282,*
Stephenson, *i, vi, 134, 167, 169, 183-184, 208, 220, 246, 273, 274-281, 285*
Stevens, Jeff, *97-98, 184-185, 208, 243, 264-265, 267-270*
Stewart, Inkleman, *63*
Stone Age Men, *3*
Striker, *42, 239-241, 246-247*

T

Tanase, Marin, *153-156*
Telescope Hill, *2, 214-215, 217*
Tennant, Colin, *45, 113, 127*
Tennis, *52, 216-217, 226*
Terman, Douglas, *32-33, 35-36, 40-43, 54, 64-65, 69, 73-77, 80-81, 83-84, 88,* 154, *226, 234, 280-281, 289, 290*
Tobago Cays, *240, 254, 258-260, 264, 137, 220*
Treaty of Aix-La-Chapelle, *12*
Treaty of Paris, *13-14*
Treaty of Utrecht of 1713, *12*
Trees
    Coconut Palm, *43, 188, 192-194, 198*
    Corkscrew, *189-190*
    Flamboyant, *169, 191, 196- 197*
    Manchineel, *9, 20, 68, 195-196*
    Sea Grape, *189-190, 192*
    Tamarind, *198*
Tref Leaves, *7*
Trio, *256*
Turtles, *119-120, 220*

U

Ubutu, *6-7*
Union Island, *iii, vii, 13, 18, 21, 31, 46, 51, 63-64, 66, 82, 84, 105, 108-109, 111, 135, 156, 162, 164, 227-228, 241, 242, 244-245, 255, 259, 262, 268*

V

Vegetables, *18, 34, 95, 120, 200-201*
Victory, (also See Goatie) *49, 52, 89, 93*
Volcanic, *2, 43, 53, 67*

W

Wakiva, 109, 242-244
Water Pearl, 261
Weapons, *4, 8, 10*
West Indies, *xiv, 3, 9, 13-15, 56, 76, 89, 128, 192, 194, 224-226*
White Island, *260*
White, Jane, *206-208*
Williams, *74, 154*
Willoughby, Lord William, *12*
Windward Island, *2, 12, 15, 23, 32, 264*
Wine Cellar, *iv, 44, 91, 99, 155-156*

Z

Zeus, *116-118, 182, 245, 255*

ABOUT THE AUTHOR

From left to right, Rosa, Lana, and Maya Shore

Lana Shore is an accomplished freelance writer and photographer who is based in Colorado. She is dedicated to conserving and preserving nature by raising awareness of the Earth's beauty through her photography and writing.

Made in the USA
Columbia, SC
01 March 2020